T0091604

SOLAR EUPHORIA

Jenny Stanford Series on Renewable Energy

Series Editor
Wolfgang Palz

Titles in the Series

Vol. 1

**Power for the World: The Emergence
of Electricity from the Sun**
Wolfgang Palz, ed.
2010
978-981-4303-37-8 (Hardcover)
978-981-4303-38-5 (eBook)

Vol. 2

**Wind Power for the World: The Rise
of Modern Wind Energy**
Preben Maegaard, Anna Krenz,
and Wolfgang Palz, eds.
2013
978-981-4364-93-5 (Hardcover)
978-981-4364-94-2 (eBook)

Vol. 3

**Wind Power for the World: International
Reviews and Developments**
Preben Maegaard, Anna Krenz,
and Wolfgang Palz, eds.
2013
978-981-4411-89-9 (Hardcover)
978-981-4411-90-5 (eBook)

Vol. 4

**Solar Power for the World: What You
Wanted to Know about Photovoltaics**
Wolfgang Palz, ed.
2013
978-981-4411-87-5 (Hardcover)
978-981-4411-88-2 (eBook)

Vol. 5

**Sun above the Horizon: Meteoric
Rise of the Solar Industry**
Peter F. Varadi
2014
978-981-4463-80-5 (Hardcover)
978-981-4613-29-3 (Paperback)
978-981-4463-81-2 (eBook)

Vol. 6

**Biomass Power for the World:
Transformations to Effective Use**
Wim van Swaaij, Sascha Kersten, and
Wolfgang Palz, eds.
2015
978-981-4613-88-0 (Hardcover)
978-981-4669-24-5 (Paperback)
978-981-4613-89-7 (eBook)

Vol. 7

**The U.S. Government & Renewable
Energy: A Winding Road**
Allan R. Hoffman
2016
978-981-4745-84-0 (Paperback)
978-981-4745-85-7 (eBook)

Vol. 8

**Sun towards High Noon: Solar Power
Transforming Our Energy Future**
Peter F. Varadi
2017
978-981-4774-17-8 (Paperback)
978-1-315-19657-2 (eBook)

Vol. 9

The Sun Is Rising in Africa and the Middle East: On the Road to a Solar Energy Future
Peter F. Varadi, Frank Wouters, and Allan R. Hoffman
2018
978-981-4774-89-5 (Paperback)
978-1-351-00732-0 (eBook)

Vol. 10

The Triumph of the Sun: The Energy of the New Century
Wolfgang Palz
2018
978-981-4800-06-8 (Hardcover),
978-0-429-48864-1 (eBook)

Vol. 11

The Triumph of the Sun in 2000–2020: How Solar Energy Conquered the World
Wolfgang Palz
2020
978-981-4800-84-6 (Hardcover)
978-1-003-00086-0 (eBook)

Vol. 12

Solar Euphoria: The Rise of Photovoltaics to the Top
Wolfgang Palz
2023
978-981-5129-00-7 (Hardcover)
978-1-003-43866-3 (eBook)

Jenny Stanford Series on Renewable Energy
Volume 12

SOLAR EUPHORIA
The Rise of Photovoltaics to the Top

Wolfgang Palz

JENNY STANFORD
PUBLISHING

Published by

Jenny Stanford Publishing Pte. Ltd.
101 Thomson Road
#06-01, United Square
Singapore 307591

Email: editorial@jennystanford.com
Web: www.jennystanford.com

British Library Cataloguing-in-Publication Data
A catalogue record for this book is available from the British Library.

Solar Euphoria: The Rise of Photovoltaics to the Top

ISBN 978-981-5129-00-7 (Hardcover)
ISBN 978-1-003-43866-3 (eBook)

Dedicated to

Jean-Marc Reiser, French Solar Pioneer.

A fresque from Egypt 3000 years ago.

Contents

*Introduction: Photovoltaics, the Sun's Victory
to the Benefit of Mankind*   xix

PART I: SOLAR POWER TODAY

1. **An Introduction to PV, to the Expert, the Non-Expert,
   for Education, for the Potential Investor**   **3**

   1.1   The Principles of Photovoltaics   3
         1.1.1   Solar Power: No Alternatives to PV   3
         1.1.2   Semiconductors for PV   4
   1.2   Ingredients for the Silicon Solar Cell   6
         1.2.1   The Poly-Silicon Feedstock   6
         1.2.2   Multi-Crystalline and Mono-Crystalline
                 Silicon Blocks and Ingots   7
         1.2.3   The Silicon Wafers   7
   1.3   Basic Features of a Modern Silicon Solar Cell   7
   1.4   Silicon Modules/Panels, Strings, and Inverters   10
         1.4.1   Global Module Industry   12

2. **PV Markets in the World Today, 2022, A TW
   Installed**   **15**

   2.1   The Revenge of the Sun: Solar PV in a
         "Rainy Country"—The Surprising Case of
         Brussels in Central Europe   16
   2.2   The Global Solar PV Markets since 2021   20
         2.2.1   PV Power in the United States   26
         2.2.2   PV Power in Europe   36
                 2.2.2.1   Largest PV markets in the EU
                           member countries   42
                 2.2.2.2   The largest European PV markets
                           outside the EU   52
         2.2.3   PV Power in China   56
         2.2.4   PV Power in Japan   61

	2.2.5	PV Power in India	63
	2.2.6	PV Power in Australia	64
	2.2.7	PV Power in Southeast Asia	65
	2.2.8	PV Power in Latin America	68
	2.2.9	PV Power in Africa and the Middle East	69

3. Climate Change and PV   73

4. Global Energy Transformation. A World of PV?   81

PART II: REVIEW OF A FASCINATING PV EXPANSION IN THE 20TH CENTURY

5. PV's Scientific Beginnings since Becquerel in Paris   91

6. 1954: Start-Up of PV's Industrial Development   97

7. US Pioneers Had a Dream   101

8. PV Front Runners in Europe   107

   8.1   European Projections for Large-Scale PV Deployment   113

   8.2   PV Programmes of the EU Paving the Way since the 1970s   115

      8.2.1 Early PV Demonstration Plants in the United States and the EU PV Pilot Programme   117

9. The Later Years of the 20th Century: Big Oil, Japan, Africa, and the Developing Countries   129

   9.1   The Oil Companies and PV   129

   9.2   Global PV Leadership of Japan at the Turn of the Century   131

   9.3   PV in Africa and the Developing Countries   132

10. PV Expert Meetings around the World, Political Monster Meetings in Support of the Renewables, New Energy Organisations   135

PART III: THE BIG BANG IN THE YEAR 2000

11. New Political Regulations in Germany, New PV
    Industry in China   143

12. 2012, PV's Year of Industrial and Social Disaster,
    Germany Giving Up Its Leadership   149

13. PV Markets up to Now: A Summary   155

14. Green Hydrogen, the Follow-Up of Solar, Wind, and
    Hydro Power   157

Index   163

Dr Wolfgang Palz

The author's portrait (from a speech given in Nanjing in 2011).

Some Milestones in the Author's Professional Life

- 1965, Univ. Karlsruhe, Germany: Diploma in Physics. PhD in PV
- 1966–1970, University Nancy, France: Maître de Conférences
- 1970–1976, Paris, CNES (French National Space Agency)

  o 1973, Paris, UNESCO, Congress organiser "The Sun in the Service of Mankind"

  o 1976, Paris, UNESCO, Book publication "Solar Electricity"

- 1977–2001, Brussels, Official, EU Commission

  o 1977–1998, Division Head, Renewable Energies R&D for Europe

  o 1977, Luxemburg, Launch of 1st EU PVSEC

  o 1993, Budapest, ISES World Congress, "Power for the World, a Global PV Action Plan"

- o 1998, EuropeAid, Coordinator EU renewable energy aid for Africa
- 1999–2002, Berlin, Bundestag German Parliament, Member "Enquête Commission" on Germany's energy by 2050
- 2003, Germany, Order of Merit
- 2004, Brussels, EU, expert contract with EuropeAid, coordinator of PV projects for the poor in Latin America
- 2006, Beijing, China, Co-organiser "1st Great Wall World Renewable Energy Conference"
- 2007–2009, Paris, ADEME (French Energy Agency) Member Scientific Board
- Swiss Solar Prize 2010
- 2010–2016, Paris, French Government Programme Investments for the Future, R&D and Innovation PIA-1, Member Commission Nationale des Aides CNA
- International Solar Energy Society ISES, Global Leadership Award in advancing Solar Policy, 2011
- 2018, Singapore, with Jenny Stanford Publishing: *The Triumph of the Sun*
- 2023, Paris, 50th Anniversary Congress, "The Sun in the Service of Mankind"

The author with Dominique Campana, Director, The French Agency for Ecological Transition (ADEME), at a restaurant in Paris where emperor Napoleon left his hat as collateral to the restaurant's owner to pay off his debts.

Wolfgang Palz: A Vision for Solar Energy since 2003

Speech by Dr Wolfgang Palz given when he received the EU Becquerel Award at the World PV Conference, in Osaka, Japan, May 2003, a Congress jointly organised by the PV communities in the United States, Asia, and Europe. The laudatio had been given by Prof. Antonio Luque, Madrid, Spain.

This 3rd World Conference on Solar Energy is again a wide forum about the science of Photovoltaics. Let me take the opportunity of this particular ceremony of the Becquerel Prize Awarding to make a few points on the wider role of solar PV in society, its future socio-political potential the way I see it.

It is my firm belief that solar energy and other renewable energies will ultimately dominate the whole energy sector worldwide. We can, in particular, expect trustfully that PV will play a central role for sustainable electricity production. Much of that PV capacity will be building integrated for decentralised generation. Since a few years PV has indeed passed the point of no return. A world without PV has become unthinkable. The milestone of 1000 MW capacity, installed worldwide, was achieved and a dynamic process is in full swing in which industry and politics invest vigorously in new plants and progressive production. The world community of PV, gathering here in Osaka can be proud of what has been achieved in a global effort since the invention of the first practical solar cell half a century ago.

Photovoltaics has, moreover, a particular fascination which goes beyond its strategic interest in the energy sector. The connection it provides to the user to the Sun reaches for some into the field of mystical experience. The Sun was worshipped in the old cultures of Egypt and pre-Columbian

America. The Emperor of China derived his power from the Sun in a dedicated temple. In the meantime, we have learned that there exist an infinite number of suns in the universe. The fact remains, nevertheless, that our particular Sun is not only the source of all our energy, it is the source of our life. Idolising it is no nonsense.

We have a world climate problem and the experts from the IPCC tell us that climate change is not only with us already; it is going to be dramatic in the course of this century. This is a tremendous incentive to go ahead fast with the deployment of the renewable energies.

Another problem is terrorism and war conflicts. My generation in Europe had the favour to live in a second half of the 20th century of peace and prosperity. But the generation of my father lived in a period of war, persecution, and misery. Since the time man has existed, the peoples make war on each other and on nature. The peace from which I was able to benefit was a local one. Africa is plagued by civil wars and their horrors.

We all know that the renewables and in particular PV are tailor made for stimulating progress in the rural areas of the developing countries, the forgotten countries where misery and poverty drive the young generations into emigration and eventually lead to social unrest and conflicts. That enormous potential of PV to change the life of 2 billion people is not taken seriously in the political world.

May I appeal to you to make your voices heard that PV has a role to play, here and now, to bring human dignity and PEACE to the world.

Introduction: Photovoltaics, the Sun's Victory to the Benefit of Mankind

Opinion polls make it currently very clear that climate change comes high up in people's concerns. No wonder with the images of forests going up in fire in many places and floods setting formerly dry areas of the Sahel or Pakistan under water. Billions of dollars or euros of cost and damages to large areas of land, their ecosystem and the people living there are arising. There is a consensus for an urgent need to do something about it.

Valuable ways and means to combat the climate getting out of control have been made and are in progress. Energy conservation, for instance, in buildings, and preservation of the ecosphere, planting new forests are widespread; recently there came even new hope for the Amazon rainforest in Brazil.

But where we are being hurt the most is the conventional energy supply. Globally, fossil fuels are still in the lead and greenhouse gas (GHG) emissions are higher than ever; even the financial support by banks and others to the polluter was never as high as it is by now.

Politicians promote, among the best energies emitting no CO_2 they know, the nuclear option. In particular so if they are owners or supporters of the atomic bomb as both are linked through the fuel cycle. Purposely omitting the fact that nuclear power is dangerous and currently the most expensive on the market, with decades of plant construction time. In addition, as shown by France, where more than half of the nuclear park of 59 plants are out of order in 2022/23, nuclear is unreliable and the government is calling on renewable power to get over the winter needs.

Some are anxious to demonstrate their commitment to solar power and invest in the concentrating solar power (CSP). CSP comes always big and politicians like it big and concentrated so they can control it, no matter if it is the most expensive among the renewable power options.

On top comes ITER, called "the most expensive science project of all times". Thirty-five governments have joined a $45 billion project in Southern France, a project to bring the Sun down to Earth. Aiming to get limitless power by imitating the nuclear fusion process in the Sun. R&D on an ITER Tokamak have already started 50 years ago, and still no fusion in sight.

Politicians may invest big sums on alternatives to fossil fuel power, but they fail to make a difference. GHG emissions and temperatures keep rising. At the same time people around the world go for the other solar energy, PV, that is already with us to alleviate climate change; and it is profitable already now for investors, small and big. PV, that's the Revenge of the SUN.

The present book is focused on solar photovoltaic (PV) power. PV is clean and sustainable: During production, installation, operation and maintenance, it emits no CO_2, other greenhouse gases, or pollutants. The saying that "the best energy is the energy you do not consume" is true no more in the case of PV. It is good and positive to go for PV as it improves living conditions and of life, but not only.

The book, written by a long-time expert on the topic, goes into some details of solar electricity, solar cells and panels and promotes a better understanding why PV is the royal champion to combat climate change and displace fossil fuels and nuclear from the world's power markets.

A new semiconductor world of PV power generation, computer chips for informatics, and LED lighting were all the three developed by a new generation of scientists, the generation of scientists after WWII, my own generation. Both PV and LED are diodes, one converting light into electricity and the other converting current into light. They are made from different materials.

From the same material, ultra-pure crystalline silicon, are produced PV on the one hand side and electronic chips on the other. The element silicon is available in unlimited quantities everywhere on Earth—unlike lithium that has become essential for the production of efficient battery storage.

PV is no more a little newcomer on the global power markets. In 2022, 1000 GW (1 TW) was already installed. It is a new record for PV: its share on the markets keeps increasing all the time while the share of power from fossil fuels stagnates or decreases like the part of atomic power. Over 16% of the world's overall power capacity installed today, all sources combined, is already PV. The potential is huge: the whole world could even be fed with PV generated electricity. Put all of the needed area to be exposed to the Sun into one spot, 440,000 km², not larger than Spain, would suffice.

Since several years already most new power installations in the United States and other important countries are PV. Most new investment finance in energy goes currently on PV.

Most important, PV is beating all others on cost. PV has become the producer of the cheapest electricity on the electric grids.

Like no other power source, PV power is decentralised. Many millions of house owners around the world seized the chance to become their own energy provider. Roof-integrated PV is amortised in 4 to 6 years. "Not investing in PV is throwing money through the window."

PV comes with the brilliant side of the Sun: the dark side of the Sun is climate change. But that one is not the Sun's fault; it is man-made. It is one of the terrible things mankind is responsible for, just to mention them, poverty, corruption, murder, and wars. The world bears ever-increasing military expenses, they come on top of state budgets around the world. Even nuclear wars are on the agenda.

PV is something peaceful, not suitable for conflicts and fighting. It improves living conditions. Its deployment in rural areas and in the less developed countries was from the beginning a dream of the PV developers. But the rich industrialised countries preferred to use it themselves first when PV entered the commercial scale by the turn of the century.

Now that PV has become a giant there is hope that everybody can benefit from it, everywhere on Earth.

PART I

SOLAR POWER TODAY

Chapter 1

An Introduction to PV, to the Expert, the Non-Expert, for Education, for the Potential Investor

1.1 The Principles of Photovoltaics

It is very simple indeed; it is just a semiconductor diode. No moving part. Delivering power and current from light incident from any direction, at any intensity. The effect is proportional to light intensity, diffuse light is accepted the same way as direct irradiation from the Sun.

The consequence is that sizable amounts solar energy can not only be obtained in Earth's solar belts but as well in the moderate climates, for instance, in the USA, Europe, or China with large populations and great energy needs. The "north façade" of a building covered with PV produces electricity as well no matter if it is less for the lower intensity there. Bi-facial cells and modules have recently become more popular as they add the power gained on the back from the albedo, the light reflected from the ground, to the power of the module's front exposed to the sunlight. Modules installed vertically are also a possibility.

1.1.1 Solar Power: No Alternatives to PV

The photovoltaic effect works best at low temperature. The effect is contrary to solar heating—employed, for instance, in water

Solar Euphoria: The Rise of Photovoltaics to the Top
Wolfgang Palz
Copyright © 2023 Wolfgang Palz
ISBN 978-981-5129-00-7 (Hardcover), 978-1-003-43866-3 (eBook)
www.jennystanford.com

heaters. When concentrating the Sun's radiation with mirrors high temperatures can be obtained, too. An example is the solar furnace at Odeillo in Southern France that can produce over 3000°C, enough to simulate nuclear explosions.

Sunlight concentration with mirrors is also employed in solar power plants. It is possible heating by strong solar irradiation a working fluid to drive an engine and a generator. In the late 20th century when PV was for many "energy specialists" a hopeless contender for solar power implementation, a lot of money was thrown after CSP. Quite a few sizable CSP plants have actually been built and operated, in the United States, Spain, Morocco, and other countries. CSP keeps fanatical adepts until today but virtually no new CSP plants are being built by now. What broke its neck was cost. Eventually CSP turned out much more expensive to build and operate than PV. Moreover, it had and has the drawback to come only in big plants that need direct solar irradiation to operate.

Being clean and renewable like CSP is excellent but PV can do better:

The PV of today is not only cheaper. It turned out to be nowadays the cheapest of all power sources, be they conventional like nuclear, coal, gas, and oil, or even renewable like the CSP or wind power. PV comes at any size, decentralised, and for any light intensity—direct or diffuse light. On top of low cost comes the advantage of short installation time, opportunities for series production in industrial factories, long lifetime and low cost of operation and maintenance. Fit for markets in all the world's climates, not only the solar belts.

1.1.2 Semiconductors for PV

There is a large choice of semiconductors suitable for PV. Preferable are those having the potential for low cost of production and operation. International development later at the 20th century went for the "thin-film" solar cells and modules. Those benefitting from industrial interest were in particular the following:

- CdTe cadmium telluride and similar compounds

- a-Si amorphous silicon
- CIS copper indium di-selenide and related ones

CdTe: The PV effect in CdTe has the highest theoretical efficiency. CdTe cells and modules were first developed in Germany by the late Dieter Bonnet[1]. They are still manufactured and commercialised today in 2022/23 by First Solar, Inc., a company quoted on the stock market. During the first boom of PV in the new century CdTe modules were mainstream on the German markets. Today they keep only a minor role with a share of some 3% of the global PV markets, mainly in the United States.

Amorphous silicon: Doping of a-Si when it has small amounts of hydrogen included was first discovered by Spear in Scotland. Early on late Hamakawa made it mainstream in PV R&D in Japan. Towards 2010 some 15% of the world's PV market was a-Si. The producer was the Japanese Sanyo. Because inherent degradation due to the high mobility of the small hydrogen ions it contains it disappeared from the markets except on the little pocket calculators. But it survived until today on the Si/a-Si hetero-junctions that are among those becoming mainstream today: those are derived from Sanyo's "Hit cell": an N-type silicon crystal that is on the top covered with an extremely thin layer of p-type amorphous silicon, just 100 atomic layers thick. Sanyo was acquired by Panasonic in 2009. The patents for the HIT cell have expired in the meantime and anybody can now employ it for building hetero-junction cells.

CIS: It is a real thin-film cell, only a few microns thick. Over 20% efficiency has been obtained in Germany. Commercialisation by Solar Frontier gave it a little 3rd place on the global thin-film PV markets until now. But recently discontinued industrially, it is said.

One may say, as a joke that in the early days the Germans had their CdTe panels, the Japanese their a-Si panels, and the Chinese their silicon panels. And now everybody uses silicon. The joke would not even be correct: Japan and Germany, like the United States, had always a strong silicon PV industry.

---

[1]*Power for the World*, Jenny Stanford Publishing. 2011, p. 165.

1.2 Ingredients for the Silicon Solar Cell

The world of PV is already since many years dominated by the silicon solar cells and modules. Today in 2022/23 some 95% of the world market is covered by silicon PV. Thanks to industrialisation to the full and mass production it is the cheapest.

Today's world market is dominated by Chinese products.

Already at the turn of the century, industry in China focused attention on silicon PV production and commercialisation. Over 80% of poly-silicon, silicon cells, and modules are imported today from China, as well as 97% of the wafers.

1.2.1 The Poly-Silicon Feedstock

Quartzite, silicon oxide, is reduced in big factories to "**metallic silicon**" at a purity of 99.99% (4N). **Poly-silicon "solar-grade" 99.9999% (6N)** up to (9N), the latter for mono-crystalline silicon production is then obtained by refinement in the classical "Siemens process" or alternatively in a fluid-bed reactor. A rather small part of it, 5% or so, goes as "**electronic-grade**" to the global chip industry, the base of today's world of informatics. For that purpose, it is further refined to hyper pure 10N and up to 13N.

One of the oldest producers of poly-silicon is Wacker Polysilicon in Germany. Today there is also Hemlock Semiconductor Corp. in the United States, REC Silicon in Norway and the United States, OCI in South Korea, GCL-Poly Energy Holdings with many more others in China: Yongxiang Polysilicon (Tongwei Group), Xinte Energy, Daqo New Energy, etc. The value of the poly-silicon market in 2022 is estimated at $US 8.5 billion for a production capacity of 1 million tonnes.

But the market has nothing of a quiet river: Chinese producers that dominate the market have ups and downs due to COVID-19 and other political restrictions and maintenance work in the factories. By end 2022 the spot price stood at $24/kg. By Jan. 2019 it had been $13/kg, but by mid-2022 it had been all the way up to $43/kg. This feedstock increase in 2022 entailed a price increase in the value chain of silicon modules that follows and eventually in the price of the Chinese silicon cells and modules.

1.2.2   Multi-Crystalline and Mono-Crystalline Silicon Blocks and Ingots

Ingot casting of multi-crystalline silicon has originally been developed in Germany by Fischer and Gauthier. Until recently a part of the solar cells on the world market were produced from this material. More details are given later.

This multi-crystalline or poly-crystalline material is further transformed into mono-crystalline silicon. For the purpose one employs traditionally the **Czochralski** process of crystal drawing from a melt. It was invented a hundred years ago by a Polish Scientist of that name. Today preference is more and more given, too, to an alternative process, the floating-zone process (FZ).

1.2.3   The Silicon Wafers

Multiple wire sawing of solid silicon blocks into poly- or mono-crystalline modules has become standard today. The process was originally developed by Charles Hauser in Switzerland. He had been formerly employed by Solarex in the United States.

1.3   Basic Features of a Modern Silicon Solar Cell

The cells have a thickness of between 150 and 200 micrometres (less than 1/5 of a millimetre). The cells are square shaped from 15.7 by 15.7 cm, up to 21 × 21 cm (called G12). The larger ones developed strongly on the markets, over 100 GW by end 2022. The wafers are phosphorus diffused N-type, with boron doping (P-type) to form a P-N junction. Light absorption on the top is improved by an anti-reflective coating after grooving it by etching (the "black cell" first developed by Lindmayer at Solarex). A back surface field introduced by a highly doped layer on the back improves performance by reducing surface recombination velocity.

The **PERC** cell introduced in the 1980s at the UNSW in Australia by Green has in addition a dielectric passivation layer on the back that reflects the non-absorbed light back into the mono-crystalline silicon to give it a second chance for absorption.

The **HJT** cell is a Hetero-JuncTion formed by a mono-crystalline wafer and a super imposed micro-thin layer of a-Si. They often come as bi-facial cells as the a-Si deposition can be done in one step.

For illustration, the details of a mainstream silicon cell are given below (It is random choice. There are a great number of companies offering similar products.) Note that this one is *not* a hetero-junction and it is not bi-facial:

From Lightway Solar in China: LWM5BB-PERC-223

Dimension: 158.75 × 158.75 mm

Cell thickness: 190 micrometres

Antireflection coating: silicon nitrite

Back Surface Field: aluminium

Maximum power (several for choice), e.g. P_{max} 5.49 W

Voltage at P_{max} 0.575 V

Current at P_{max} 9.555 A

Cell efficiency 21.8%

Fill Factor 80.64%

Temperature coefficient for P_{max} −0.38%/°C

Price from €0.172/Wp

In 2022, the race is on among the leading Chinese manufacturers for higher sophistication of the silicon cell process.

Firstly, the trend went from poly-crystalline PERC to mono-c P-type PERC, the former having almost disappeared from the markets by now.

Further sophistication comes by upgrading the latter by the "POPCon process" on N-type phosphorus-doped silicon. Tunnel Oxide Passivated **POPCon** was developed by FhG ISE in Freiburg, Germany since 2013; it was first used since 2019 in China on bi-facial cells, alternatively to the HJT process. The principle is to add a nano-thin oxide layer suitable enough for tunnelling on the cell's back. It allows harvesting more energy from the rear side compared to PERC bi-facial modules. Jinko, Longi and others announced efficiencies above 25% with it. In October 2022 JinkoSolar proudly reported to have achieved in research a large-area mono-crystalline cell of N-type of an efficiency of 26.1%. Trina began mass production of its Vertex N cells in March 2022 with 210 mm cells. Trina achieved a top of 25.5% efficiency. The principle of POPCon as a layer-structure is given below:

SiN_x top coating

- Passivation layer
- Doped P^+ emitter
- N-type silicon crystal
- Ultra-thin tunnelling layer
- Doped N^+ silicon
- SiN_x rear coating

As a third step one can go further and apply Interdigitated Back Contact (**IBC**). IBC applied silicon cells have no visible grid lines on the front anymore, both contacts are buried on the back. IBC was first proposed in 1995 in the United States by Verlinden and Swanson at SunPower.

The **efficiency** of a commercial large-area mono-crystalline cell nowadays comes between 21 and 23%. It is a bit less than that, rather 18% or so on a poly-c silicon cell.

The International Technology Roadmap for Photovoltaics (ITRPV) from the German VDMA published its 13th edition in April 2022 based on inputs from 62 international experts:

It foresees a higher-grade N-type silicon share increasing from a current 20% to 70% in the next ten years. The share of mono-crystalline material is 90% in 2022 and will be further increasing. *As a personal note I mention here a comment I got in the 1990s from people of BP Solar in England, a leading cell manufacturer at the time after I had underlined the jewel-like glittering of the multi-crystalline cells: Tim Bruton of BP solar then said he agreed with all my paper, but the cells had to be mono-crystalline and black. Today they are black indeed.*

ITRPV found furthermore that 85% of silicon cells employ PERC technology in 2022; its share will slowly decrease to 70% by 2032.

Hetero-junctions are going to reach a 19% market share.

Smaller wafer formats of 161 mm (M6) will disappear in the next 3 years with a new focus on 182 mm (M10) and 210 mm (G12), they say.

In 2021 we had contrary to all the preceding years no price reduction on the global silicon cell markets because of the tripling of the poly-silicon price. By the end of 2022 the prices of the poly, the cells and modules had stabilised again and even lowered as in all the years before.

1.4   Silicon Modules/Panels, Strings, and Inverters

The power of a module is given in **watt$_{peak}$**, or simply W in our examples here below. It applies for an incident radiation of 1 kW per m^2 at sea level, the maximum available after light absorption in Earth's atmosphere. Unless the module is oriented to follow the apparent move of the Sun, this power is always lower throughout the **8760 hours** of the year. As a rule, one may say that a module of 1 kW generates a yearly electricity of 1500 kWh ± 25% depending on the location, its climate, and the module mounting. The world's highest has been reported for large productive plants in California: 2800 kWh/kW in a year.

60 cells, 72 cells and recently even 132 cells are connected into a module. Only in 2019 PV industry had passed the 400 W that a year later a strong trend goes towards 1 kW per silicon panel. Many of the key producers offer now panels over 500 W or 600 W; JA Solar broke the record with 800 W.

The mainstream module producers have the trend to employ 3 or 4 new technologies at the same time: for instance, PERC+bi-facial+halve-cell.

The half-cell technology was developed in 2014 by **FhG CSP**, Center for Silicon PV, in Halle, Germany: Researchers cut 72 square cells into 144 half cells and got 5% more efficiency compared to the full cell module. The voltage of the module is double and the current half and the power the same: the lowering of the current half means lowering half the loss by the electric resistance of the cell.

ITRPV writes that the introduction of half-cell technology and **larger-format modules** have established themselves by now. For power plant applications 60% of the modules are larger than 2.50 m² today; that share is expected to reach over 85% by 2032. 16% of the modules are larger than 3 m². On roof tops 50% of the modules are 1.80 m² and less. As a reminder, 2 m² at 20% efficiency would mean 400 W_{peak}. In 2021, mono-crystalline modules came at an efficiency of 20% or so, the multi-crystalline ones rather at 17.5%, the CdTe modules 18%.

For illustration we go in the following to some details of the modules to be currently installed in the Estremadura in Spain. With 379 MW and 588,000 modules the project is significant:

From Trina Solar the TSM-DEG 21C.20

The power of a module is 640 W

Efficiency 20.6%

37.3 V_{mpp} (mpp = at maximum power point)

17.2 A_{mpp}

132 mono-crystalline cells

Bi-facial,

Half cells

Temperature coefficient −0.34%/°C

Encapsulation double glass 2 mm

Weight 38.7 kg

Twenty-eight of those modules are connected into a string. The strings go into **MPPT** (Maximum Power Point Tracking) inverters of 60 A at 1500 V. To take into account some losses due to conversion in the inverter one may then give the power P_{ac} from the input power P_{dc}. As a rule, P_{ac} comes some 20% lower than the P_{dc}.

The output of a PV panel comes always as a DC current, polarity + and –. Inverters are employed to convert this one to AC current suitable for the national grids. The grid in Europe operates at 50 Hz or sine waves per second, in the United States it is 60 Hz. The grids keep the frequency very stable and constant, more so than the power in the various parts of the grid.

1.4.1 Global Module Industry

Today's major module manufacturers produce all silicon modules, except First Solar. The leading ones on the global markets are the following, including some manufacturing sites (2021 ranking, compiled by **RTS Corporation** in Japan):

- LONGI Green Energy Technology, **China**, Malaysia, Vietnam
- Trina Solar, **China**, Thailand, Vietnam
- JA Solar Technology, **China**, Malaysia, Vietnam
- Jinko Solar, **China**, Malaysia, USA
- Canadian Solar, CANADA, **China**, Brazil, Vietnam, Thailand, Indonesia
- Hanwha Solutions, **Korea**, China, Malaysia, USA
- Risen Energy, **China**, Malaysia
- First Solar, **USA**, Malaysia, Vietnam
- Chint Electrics, **China**, Thailand, Vietnam
- Wuxi Suntech Power, **China**, Indonesia

The number 1 LONGI had in 2021 a shipment of PV modules of 38 GW, twice as much as in 2020. In 2 years, it expects doubling again production capacity to over 100 GW. First Solar with CdTe had in 2022 a module production capacity of 9.5 GW. So far, the report of RTS in Japan.

But in the global module business, too, competition is harsh. Above ranking is seen rather differently in the following list, also for 2021 (given in the "Spring 2022 Solar Industry Update" published in April 2022 by **NREL** at Golden, Col. Authors D. Feldman, K. Dummit, J. Zuboy, J. Heeter, Kaifeng Xu, R. Mergolis. We will also come back later to this important report)

Aiko	31.2 GW shipment
Tongwei	27.3 GW
Longi	19.6 GW
Jinko Solar	16.9 GW
Zhongli Talsun	10.7 GW
JA Solar	10.7 GW
Canadian Sol.	8.7 GW
Runergy	8 GW
Hanwha Q-Cells	7.6 GW
First Solar	7.6 GW

The authors note that in the 5 years from 2016 to 2021 shipments of the 10 majors grew from 33 GW to 148 GW. In 2016 the number one had been **Trina** with "only" 5 GW; in 2021 it even disappeared from the list of the 10 majors. The discrepancies between both reports, both from very reliable institutions, is still surprising, and as far as Longi is concerned, the American listing is certainly wrong as we know from still other sources. Trina's is not correctly reported either. As a matter of fact, Trina shipped in 2022 over 40 GW. It was even no. 1 of the world's 210 mm square cells. Its Vertex N cells got in production in 2022.

The NREL notes that new companies moved quickly to the top in part through the rapid growth of mono-crystalline production. This may make sense as the share of mono-c silicon on the global markets exploded, but like in the case of Trina it overlooks the fast reaction of some Chinese manufacturers.

Chapter 2

PV Markets in the World Today, 2022, A TW Installed

Flower carpet in Brussels.

Solar Euphoria: The Rise of Photovoltaics to the Top
Wolfgang Palz
Copyright © 2023 Wolfgang Palz
ISBN 978-981-5129-00-7 (Hardcover), 978-1-003-43866-3 (eBook)
www.jennystanford.com

2.1 The Revenge of the Sun: Solar PV in a "Rainy Country"—The Surprising Case of Brussels in Central Europe

Before addressing recent global successes of solar power on a very large scale and worldwide, and go into their many facts and figures, the case of a "not very sunny" city is intriguing and merits attention as it shows the practical value of solar PV for just any climate. Here comes the story of PV in **Brussels**.

Obviously, Sun seekers in Europe may go to Spain or the Côte d'Azure but not to Brussels. The climate and living conditions in the city are not really bad, but solar radiation is not abundant. And still, solar PV has a great success story there.

Brussels has not more than a million inhabitants. It is the capital of Europe, 27 member countries forming the European Union, the EU, now without the UK. It houses the EU Commission, the EU Council, and the European Parliament, all occupying some 50 buildings. This implies some 20,000 expats plus their families and another 100,000 of lobbyists. Brussels benefits yearly with some €5 billion from the presence of the EU institutions.

The NATO headquarters is also in Brussels since De Gaulle kicked it out of France. Brussels is very international, with immigration from many countries. It has an important Islamic community, too. Not long ago it had terrorists throwing bombs in the airport and a metro station in the European quarter (with many victims)—I used that one often myself usually working in the area, fortunately not that day.

Brussels is also the capital of Belgium. It has a king that resides in Brussels. Belgium exists only since 1830 but Brussels has a long prestigious history. In the 16th century it was the capital of Charles V, (Charles Quint in French) the great "Holy Roman Emperor", of an "Empire on which the Sun never sets". He was the head of Spain when it conquered America and at the same time the Emperor of Germany when the country was unsettled by Luther and his protestant supporters, the beginnings of a religious division of the country and its later ruin in the terrible 30-years war.

Charles V was a Habsburger, but he was born and raised in Flanders, he was a Flame. Brussels, his capital city, was Flemish and became bi-lingual Flemish-French only in the 20th century with French as the common speech today.

Today in 2022 the revenge of solar PV in Brussels is particularly impressive, an explosion of demand, installers can hardly follow. I installed in the year 2000 Brussels' first kW of PV on my house there. At the beginning of 2022 the city had 12,000 PV installations with 207 MW in total. Most of them had been installed free of charge for the owner. Later more on the PV on roof tops in Brussels for free.

Inauguration of the author's PV house, the first in Brussels, from the year 2000. On the left is the mayor of the town.

(*Some details for the reader interested in financing*: In Brussels you may install on your roof, for instance, 5 kW: in 2022 they come 1 to 2 euros/W in Brussels, say a mean €7000 for your 5 kW. Given the latitude of Brussels you will install your panel ideally inclined 35% on the roof. For an average sunshine you would expect 5000 kWh a year (as said earlier 1000 kWh/per kW installed). Irradiation being less good in Brussels, you may collect only 4000 kWh. On the Brussels market, 1 kWh comes currently at 50 cents. Hence your installation produces €2000 annually. The average household consumption per year in Brussels being 3000 kWh; you could have more than 100% auto-consumption, but you will need a battery to bridge the intermittency of incident irradiation. In practice, as long as you have no storage, you might use 1/3 in auto-consumption and sell the rest back to the grid. If you got a variable price contract with your supplier, you would get 20 cents back per kWh sold to him. For 1/3 of the production of 4000 kWh having not to buy at 50 cents on the market because you produce it yourself you save €660 in the year. In addition, you got an income for two-thirds of the 4000 kWh collected and then sold to the grid, that is €530. The gain net in total is €1200.)

Your PV investment of €7000 is written off in less than 6 years. For the rest of the panel's life of over 30 years you get your electricity for free.

In case you add a battery to your system you may increase auto-consumption, for instance, from 30% to 70%. A rule of thumb would be to employ a battery of 1 kWh per kW of power for storage. For Brussels it is calculated that amortisation time of 6 years or so is also achieved with battery as self-consumption is increased and it is more profitable.

But you can even do better: in Brussels a PV roof owner can benefit from "**Green Certificates**" for 10 years. For plants of less than 5 kW you get 2.7 Certificates per kWh. There is a free market for them where they are valued some €90 per certificate. For your 4000 kWh plant you get up to €1000 a year. This way you can write off your plant's cost just with the Green Certificates.

A company hence can offer you an installation of a PV plant on your rooftop, totally free of charge. You can use immediately the electricity from your array. The investor gets the Green Certificates, they are sufficient for a return on his investment in 10 years. Needless to say, this scheme works only for panel prices low enough as in our example above. However, the truth is that in many countries they are much higher on the local PV market.

Belgium is a small country, but it has three languages (this must be striking for an American of a big country that has only one. And the European Union has even 24 official languages for 27 member countries). And in Belgium's three major regions each one has its own type of support for PV.

Wallonia next to Brussels did away with Green Certificates, it got **net metering** that Brussels does not have. To compensate for the use of the grid by the PV owner, the "prosumer", has to pay a tariff that can be fixed or proportional to the electricity sold to the grid. In the latter case the prosumer needs a new "net metering" meter. Battery storage in the case of disposing of net metering is not interesting, as the grid serves as the storage.

Like in many countries by now, community systems are becoming popular in Wallonia, too. In little Hannut there the electricity from PV installed on the community buildings will not be sold to the grid but for a preferential price to the villagers, a kind of power purchase agreement (PPA).

Flanders is the most advanced region for PV in Belgium. Already in early 2021 it had 500,000 rooftop installations and in total 4.4 GW installed. Prosumers get no Green Certificates anymore. Flanders has "the distribution tariffs". It is kind of net metering: you got to install a digital meter counting separately the current going in and out. For the current you sell to the grid you get less than for the one that you buy. To encourage auto-consumption you get a premium adding a battery. Currently half of new PV systems installed in Flanders come already with a battery.

For curiosity we can also mention the company Looop in Japan. It is offering nationwide Zero Yen installations. That is reported by RTS Corp. in Tokyo. The idea is somewhat different from the case of Brussels. In Japan the installer takes the ownership for ten years and the user has to pay a fee to use it. We saw that in Brussels instead the ownership does not change and the rooftop owner with PV on it can use the current produced freely from the beginning. The Green Certificates make the difference.

Similarly, and in Japan, too, Hanwha Q-Cells Japan offering in 2022 a PV system leasing plan with no initial cost for newly built houses; with a contract term of 15 years.

Not to forget that in California Third-Party owned PV, that means indeed leasing, dominates the market since more than 15 years already.

2.2 The Global Solar PV Markets since 2021

In April 2022, the world had installed 1 TW of PV solar power (1000 GW or 1 million MW), pollution free, without greenhouse gas (GHG) emissions in production, installation, operation and maintenance. It is expected that up to the end of this decade, an additional 2 TW (terawatts) of global PV will follow. What an important achievement for combating climate change on the royal route of solar energy implementation.

Since the turn of the century in 2000 when the world had in total a meagre 1 GW of PV installed, its spreading in the markets accelerated all the time. Eventually in 2022 for the first time 210 GW of PV were newly installed in one single year. The race goes on, in 2023 even accumulated 1.3 TW of PV are achieved.

In the course of 2021, a world total of 939 GW_{dc} was reached with the addition of 172 GW_{dc} installed. At 194 GW, the global shipment had even been higher. By now, more than a billion PV panels are installed worldwide. Following the NREL report, by Feldman and others, mentioned earlier, 95% of PV shipments in 2021 were mono-c silicon. In 2015, just 6 years earlier it had been 35%, with multi-c silicon at 58%; in 2021 that one has virtually disappeared from the markets.

With 76% market share, mono P-type PERC was now the dominant cell type, but N-type PERC kept growing to 6%; (up to 30 GW of it were foreseen by other sources in 2023).

TOPCon (N) had a market share of 9%, exceeding HJT/HIT silicon at 4%.

Thin-film CdTe, mainly produced and sold in the USA had a global market share of 4%.

The selling price of mono-c silicon modules on the global markets in 2021 was on average 38 US cents/W (in China it was only half of it). In 2015 it had been 60 cents and more.

Following the NREL report, global PV module prices follow an experience curve, they call Swanson's law, after **Dick Swanson** from **SunPower**. It describes in logarithmic form the average module's selling price over the cumulative global shipments. For every doubling of cumulative shipments there has been since 1976 until today a 23% reduction in the price. Since 2008 prices have declined faster however: in the experience curve global module prices should stand now at 48 cents/W, more than the 38 cents that were achieved. 38 cents/W coming for **bi-facial**, high efficiency modules. In October 2022 "Bridge to India" reported some respite for module prices. After shooting up by 65% over 2 years to $0.30/W international prices came down $0.26/W on CIF basis for delivery in Q2 2023, they say. This was expected because of massive expansion across the value chain in China.

The engine of growth of global PV markets was and is China. Not only its home market is the world's no. 1, but China is also the key supplier of PV panels to countries throughout the globe.

In 2021 China came on top of the nations both for total installation and yearly deployment in that single year (the latter in parenthesis on the list below:)

China	309 GW (+54.9 GW)
EU	167.5 GW (+26.5 GW)
USA	120 GW (+23.6 GW)
Japan	78 GW (+6.5 GW)
India	60 GW (+13 GW)
Germany	59 GW (+5.3 GW)
Australia	25 GW (+4.6 GW)
Italy	23 GW (+1 GW)
South Korea	20 GW (+4.2 GW)
Spain	19 GW (+4.9 GW)
Vietnam	17 GW (+2 GW)
Brazil	15 GW (+5.5 GW)
NL	14.3 GW (+3.3 GW)
France	14 GW (+3.4 GW)
UK	14 GW (+0.7 GW)
Turkey	7.8 GW (+1.2 GW)

(Continued)

(*Continued*)

Poland	7.6 GW (+3.7 GW)
Belgium	7 GW (+0.9 GW)
Mexico	7 GW (+0.8 GW)
Chile	4.3 GW (+0.7 GW)
South Africa	2.9 GW
UAE	2 GW
Morocco, Egypt	Mostly solar CSP, big ideas on PV

The author's speech at the Great Wall Conference, Zhangjiakou in Nov 2019, just before the COVID-19 was discovered in China"

In 2022, global PV markets accelerated further. Globally installed were more than 210 GW$_{dc}$:

China	87 GW (+60% y/y) (China installed newly in the year more PV than any other power source)
EU	41 GW (+47% y/y)
USA	18.6 GW
India	17 GW
Brazil	12 GW
Germany	7 GW
Japan	6.5 GW

In a business-as-usual scenario the IEA in Paris projects for 2027 a total worldwide PV installation of 2.36 TW surpassing for the first time that of coal.

The Arab countries are catching up fast. An example is the 800 MW PV plant at Al Kharsaah that was inaugurated in October 2022 in **Qatar**. It was built with the French Total Energies on 1000 ha of land. It has 2 million bi-facial modules mounted on 1-axis trackers. The electricity produced is sold locally by a **PPA**.

In the early days, the North African countries and the Middle East had focused attention on CSP, the solar power employing mirror concentration. By now it is understood there, too, that the better way to go is PV. Together with **CSP** the Germans had encouraged these countries also to export the new solar electricity to Germany, the **DESERTEC** project. This idea never realised either.

Tracking on utility-scale systems has become predominant in recent years. In the United States, for instance, in 2020 90% of the GWs of utility-scale systems there employ tracking; in 2014 it had only been half.

Just half of new PV power deployed worldwide in 2021 was utility scale with individual systems over 1 MW in size, 28% came for residential applications and 19% for commercial and industrial ones (IEA figures). And new ways of implementation are emerging next to building-integrated and ground-mounted PV: just to mention "agri-PV", integrating panels in agricultural cultivations, or floating PV on water...

Global PV installations in 2021 dominated all other power additions, renewable power like wind, and all conventional power like coal, gas or nuclear power, too. Half of all the world's power extension that year was solar PV! Wind power additions came ¼ of the total and new fossil fuel power stations 14% (Bloomberg).

In China solar PV provided 30% of new power in 2021. In the United States, the world's largest economy, 44% of all new power capacity in this year was PV. It ranked already first in new capacity additions in each of the last 9 years (**SEIA/ Wood Mackenzie** 2022).

Solar PV is nowadays the cheapest of all electricity providers.

For 2021, the global weighted average cost of new utility-scale PV power stood at \$857/kW while it was \$4800 in 2010. For onshore wind it came \$1325/kW and for offshore wind \$2858/kW. In energy terms, levelised cost (LCOE) per kWh came then 3.3 cents for the lowest-cost PV and 4.8 cents/kWh for onshore wind electricity.

Bloomberg NEF found that also in comparison with conventional electricity, for bulk power in 2022 kWh from utility-scale PV came lower than all others, in orders of cost:

- Offshore wind
- Natural gas
- Coal
- Onshore wind
- Fixed-axis and tracking ground-mounted PV

We will see later that in Europe owners of PV plants are now being taxed for their "excessive" income, a side effect of price increases on all grids due to the war in Ukraine. PV owners are indeed protected from cost increases on the grid for the part they don't have to buy being auto-producers. Gone the time of support needed for triggering mass production of PV.

On the other side, subsidies of billions of dollars and euros continue to flow into fossil fuel and nuclear generation. More details later in a following chapter.

Investors in energy got the message by now: in 2022 investment in PV was the highest of all times. At **over \$220 billion** PV got the double of investments in coal, gas, and nuclear power taken together (IEA figures). It was also 30% higher than that in wind power, PV's twin for the new renewable energies. It was four times the investments in classical renewable hydro, and even 50 times the one in battery storage.

Globally, each one, **utility-scale and distributed PV** passed the \$100 billion mark of investment.

For the adepts of **nuclear** power, one may add here that in the same year \$24 billion were invested in atomic energy, almost 10 times less than in PV. Six new plants have been grid connected in 2021, three of them in China, while eight have been closed.

Similarly to the 1980s and 1990s, when the oil majors were promoters of PV, they moved by now into PV again, too. The

French Total Energies mentioned before bought the PV company Solar Direct for €1.4 billion, and **ENGIE** the battery specialist Saft for €1 billion. The French ENGIE_has a total PV capacity of 4.2 GW worldwide running. Shell is active in Italy and agreed with others to deploy 2 GW in Spain. BP bought end of 2021 5.4 GW of CdTe modules from First Solar. The Italian ENI moved into the Greek market and it acquired 1.2 GW for Spain.

And there were the many other big operators in the global PV markets:

The **power utilities** that were reluctant for PV in the early days came in very strongly. ENEL Green Power owns by now over 8 GW of capacity in the world. EDF Renewable Energy, a daughter of the state-owned EDF in France that is virtually bankrupt and holds over 30 nuclear plants that are in trouble and non-operating, this subsidiary is doing extremely well with PV and other renewables. It is operating in all parts of the world, less though in its home country. EDF RE holds several large PV plants in California that are very profitable.

It is impossible to mention all of today's investors and operators in PV, banks, insurance companies, and other financial investors.

And there are the millions of private investors those who invest in distributed PV, in particular the buildings.

PV is an engine of job creation: In 2021 4.3 million people were employed in PV. Growth of jobs in PV is the highest among all renewable energies. Following **Irena** in Abu Dhabi, two-thirds of the jobs are occupied in Asia. The share of women keeps increasing.

Thus far we had concentrated interest mostly on PV POWER investments and markets. In the following we shall shortly address electricity GENERATION from that power.

In total, 939 GW had been installed globally at the end of 2021. For converting to ac power, we divide by 1.3. The 770 GW_{ac} generated some 1000 TWh of electricity on the grids, 1.3 TWh per GW, on global average.

In 2021 the world produced from PV 5% of the electricity it consumed in total. The highest share was in California with 25%. In China it was similar to the global average 5%. In the USA it was 4%, in Turkey 6%, in the EU 7%, in India 8%, in Japan 9%, and in Germany 10% of electricity consumption.

Following Ember in London we can see the following trends in electricity demand for the year 2022:

- Demand increased everywhere in the world except in the EU where it slightly decreased.
- There is nothing of a nuclear revival. Delivery of nuclear electricity declined on a world average. It slowed in particular in Europe and a bit in the United States and India. It had a minimal increase in China that is by some considered wrongly the new nuclear El Dorado.
- Supply of electricity from coal came down in the United States and even more so in China! Its consumption went strongly up in India and not a little even in Europe to compensate the gas import restrictions from Russia.
- The renewables are by now the key market players: New electricity supply from PV among the renewables is strongest in India with its recent exploding PV markets. Otherwise, it is electricity from wind power that has a longer record of installations and benefits from longer running time over the year than PV. New electricity from wind power, hydropower, and PV came roughly up for all of the increased demand.

2.2.1 PV Power in the United States

The United States is the world's economic power no. 1. How PV markets develop there is also relevant for the rest of the world. And 2021/22 was a year of Glory for PV in the United States.

The US Energy Information Administration (**EIA**) sees **23.6 GW$_{dc}$ (18.5 GW$_{ac}$)** of new PV capacity installed in 2021 in the country. PV comes on top before 11.2 GW of new wind power capacity, 9.2 GW of power from natural gas and 6.2 GW of battery storage.

Solar PV accounted for some 44% of all new capacity on the US grids.

In total, in 2021, 119.7 GW$_{dc}$ or 92.5 GW$_{ac}$ of cumulated PV capacity is installed in the country. The electricity generated from the PV power was 163 TWh in 2021.

Something New Under the Sun. It's the Bell Solar Battery, made of thin discs of specially treated silicon, an ingredient of common sand. It converts the sun's rays directly into usable amounts of electricity. Simple and trouble-free. (The storage batteries beside the solar battery store up its electricity for night use.)

Bell System Solar Battery Converts Sun's Rays into Electricity!

Bell Telephone Laboratories invention has great possibilities for telephone service and for all mankind

Ever since Archimedes, men have been searching for the secret of the sun.

For it is known that the same kindly rays that help the flowers and the grains and the fruits to grow also send us almost limitless power. It is nearly as much every three days as in all known reserves of coal, oil and uranium.

If this energy could be put to use – there would be enough to turn every wheel and light every lamp that mankind would ever need.

The dream of ages has been brought closer by the Bell System Solar Battery. It was invented at the Bell Telephone Laboratories after long research and first announced in 1954. Since then its efficiency has been doubled and its usefulness extended.

There's still much to be done before the battery's possibilities in telephony and for other uses are fully developed. But a good and pioneering start has been made.

The progress so far is like the opening of a door through which we can glimpse exciting new things for the future. Great benefits for telephone users and for all mankind may come from this forward step in putting the energy of the sun to practical use.

BELL TELEPHONE SYSTEM

Discovering PV in 1954 in the United States.

In 2022, PV reached in the United States a capacity of 140 GW$_{dc}$ with the addition of 18.6 GW, slightly less than in 2021. In 2022 California came again as top state for PV installations against

Texas, the top of the previous year. SEIA, the Solar Industries Association sees a higher increase of up to 40 GW per year from 2024 onward.

Utility-scale PV accounted for the largest part of the new PV capacity installed in 2021. By 2027 even 162 GW of it are supposed to be newly added, and we'll see below the reasons for this optimism. But at the beginning of 2022 projects have been delayed in particular through an investigation of the Department of Commerce DOC for alleged anti-dumping violations by four Southeast Asian countries importing PV into the United States. Following EIA, 20% of utility-scale projects were delayed. Most of the projects to come online in the next 18 months are already under construction, it found.

Following SEIA, the national **average system price** for utility-scale PV came $0.92/W in 2022 when it was fixed tilt and $1.07/W for tracking. For residential PV the turnkey installed price in the United States was $3.25/W in 2022, including a lot of "soft" cost (permitting, customer acquisition, overhead, margins, taxes, etc.) and only little for the price of the PV modules, the key of the system.

Uncertainties following the DOC investigation ended in June 2022 when President Biden created a 24-month **moratorium on solar tariffs** related to the investigation. In addition, on August 16 the same year, the President signed into law the Inflation Reduction Act (IRA).

IRA brought a gigantic improvement of support for PV.

There was PV in America before IRA, and there is a new age for PV after it was approved by Congress in a harsh battle within a big package of various supports and signed by the President, a masterpiece of success for him.

IRA brought considerable incentives for PV on the federal level (later we will also review more others on the regional and local level):

- On top of the new support comes a long-term extension of the **Investment Tax Credits (ITC) at 30%:** It applied already for 2022 and lasts until 2032! It applies to business projects and residential projects, too. A big support indeed!

Interconnection costs are included in the tax credits. To the 30% tax credit, an additional 10% credit can be obtained by purchasing domestically produced hardware, the panels, inverters, electric gears. Projects located in former "energy communities", brown fields and others, can get another 10% tax credit. PV projects that sell the electricity to low-income consumers get 10% tax credit more. In total one could achieve 60% tax credit.

At the end of 2021 the United States had 3,012 million residential solar systems, a number rapidly increasing (7 kW on average). In total, PV residential power increased by 2.9 GW in 2020 to +4.2 GW in 2021 and +5.6 GW in 2022.

In the United States, Wood Mackenzie keeps a "Leaderboard" of the PV installers in the country. They were 30% more in 22 than the preceding year. Sunrun was the lead installer with 13% of the total residential US market. Together with the second, Titan Solar Power and the third Freedom Forever it has even 20% of the US market. They have similar business models in the form of Third-Party sales. Tesla has been displaced in 2022 from the top three of installer ranking.

Large utility-scale PV projects (more than 1 MW capacity) can also opt for the **ITC instead of the PTC.** For getting the full 30% tax credit, laborers and mechanics installing utility-scale PV systems must be paid prevailing wages and be part of an electrical apprenticeship program.

- IRA, the Inflation Reduction Act provides alternatively the Production Tax Credit (PTC) as an option of support for PV projects.

While residential and other building-integrated PV projects with their relatively high installation cost are supposed to opt for the upfront investment tax credits ITC, the PTC is also open to them. But they will be better off with the ITC.

The large utility-scale projects in general will better fare with the PTC as their investment cost per kW is relatively low: the PTC for PV is basically 0.30 cents/kWh. But if a project meets the prevailing wages requirements, it receives for 10 years a PTC of 1.5 cents/kWh, inflation adjusted by the IRS it comes in 2022 to 2.6 cents/kWh.

- After 2023, PV will benefit from a new federal tax credit that was introduced with the IRA: the Clean Energy Production Tax Credit will replace the traditional PTC for systems placed in service 2024 or later. It applies to all facilities with zero GHG emissions.

Tax credits and depreciation benefits for solar and other projects are exempted from a new tax, **the Corporate AMT**. That one was introduced with the IRA in August 2022. It imposes a 15% corporate minimum income tax on certain large corporations intended to help pay the healthcare and climate change priorities in the IRA. It is established after international consensus on a global minimum tax regime.

- And there is **Solar Energy Manufacturing for America (SEMA)**. Through IRA the US solar industry gets access to Production Tax Credits and Investment Tax Credits for domestic manufacturing across the PV value chain. SEIA hailed in particular this key provision and the many commitments initiated thereby to build new PV manufacturing facilities domestically in the United States.

Without the effects of the new IRA legislation, PV in America is impressive already. In the following is given part of some takeaways for the reader put together by SEIA for 2021:

- 119.7 GW_{dc} of PV were in 2021 installed in the United States, 11% of the US total electric capacity of over 1200 GW.
- 3,659,000 PV systems are installed, most of them decentralised.
- 255,000 jobs are currently in the PV industry.
- There are over 10,000 PV businesses in the United States.
- The value of the US solar market was $33 billion in 2021.
- Through PV GHG emissions are reduced by 146 million tons a year.

For the future America's PV industry sees

- 30% of US electricity generation from PV by 2030 (today it is 4%; it grew on average by 33%/annum in the last 10 years).

- 480 GW of PV to be installed the next 10 years.

In the following are given some more details for the year 2021. They are compiled by information from the US agencies NREL, SEIA, and EIA.

- In 2021 the United States installed 23.6 GW_{dc} of new PV. Taking into account the conversion losses from dc to the ac of the electric grids it was 18.6 GW_{ac}.
- In the 5 years since 2017 PV has presented some 35% of all new power capacity.
- Even better, in 2021 a major part of 44% of all new power capacity installed in the United States was PV. Wind power came behind at 33%. 70% of the new PV power installed in the year was utility scale (1 MW+). The rest of 30% was distributed PV, mostly residential.
- Six states installed over 1 GW of new PV in 2021. Texas came on top with 4.3 GW followed by California and Florida.
- Many states took actions concerning distributed PV, successor tariffs to net metering, implementation of rebates, low-income programs, etc.
- In total at the end of the year, there were 119.7 GW_{dc} or 92.5 GW_{ac} of PV connected to the grids. 59.5 GW_{ac} are utility scale, 21 GW are residential, and 11.9 GW commercial and industrial.
- On top of the states in terms of cumulative PV capacity installed at the end of 2021 came California with 26.6 GW_{ac}, followed by Texas, Florida, North Carolina and Arizona. 19 states had more than a GW_{ac} of PV installed.
- For PV per capita, Nevada came first with almost 1.4 kW per person. In California it was 600 W_{ac}. For the United States on average it came some 300 W.
- The United States generated 4% of all its electricity from PV in 2021. 11 states produced over 6%. For five states it was even over 16% of PV electricity with California with 25% on top, followed by Massachusetts, Nevada, Hawaii and Vermont.
- My friend **Neville Williams**, a long-time PV pioneer wrote me that Florida has also a first "solar town" at Babcock Ranch

with a 75 MW solar plant and 10 MW of battery storage. The utilities in the state build thousands of MWs on solar farms since they know that PV is the cheapest form of energy there is.

- 4.8 GW of PV were produced in 2021 domestically, 11% more than the year before. The increase was due to 25% increase of production by First Solar; First Solar produced 1.8 GW, the world's only thin-film modules producer, CdTe modules as it were.

- 23.6 GW of modules were imported in the year. 90% of them were c-Si modules and cells. Most of them came from Malaysia, Vietnam, Thailand, and South Korea. Imports decreased 18% in 2021 from the previous year.

- More than half of the modules did not report a duty, likely because of an exemption from "Section 201" tariffs for bi-facial modules. The "Section 201" tariffs were extended for 4 years in February 2022 in a reduced version. They apply on crystalline cells and modules at 15% (in previous years as much as 30%), the first 5 GW (in previous years only 2.5 GW) of imported cells and modules are exempted as well as all bi-facial modules, as mentioned above.

- 95% of PV shipments were mono-c-Si compared to 35% in 2015 when multi c-Si peaked at 58%. Mono P PERC was dominant, but N-type is rapidly gaining market shares.

- The average module efficiency installed in the United States was 20% or so for mono-c-Si with an increasing trend, and 18.2 for CdTe. In 2010 it had respectively only been 14.2% for c-Si and 11% for CdTe: a world of technical progress.

- The average price of a PV module in the USA, before tariffs, dropped from 39 cent/W in 2018 to 27 cent/W at the end of 2021.

- National revenue from PV deployment, sales and installation of full systems, was $30 billion in 2021. It was approximately 17× higher than the revenue from module manufacturing.

- The median price for 166 utility-scale PV plants owned by 25 regulated utilities was $1.20/W$_{ac}$ or $0.97/W$_{dc}$.

- Median gross cost of a residential PV system came $2.50/W at year end 2021. As reported by EnergySage prices vary by state. In "rich" Massachusetts it may come $3/W against $2.25/W in Arizona.
- In 2021 the United States installed 10.6 GWh or 3.6 GW of energy storage, three times more than the previous year. Almost two-thirds of it was installed in California, with record levels of residential and "front-of-the-meter" deployment. But most residential PV comes without storage.

  By 2026 it is expected to mount to 57 GWh. In 2022, total storage capacity came already up to 13.4 GWh, some 10% of it for residential applications, mostly in California and Texas.
- Residential PV+battery could have a total price of $5400/kW$_{ac}$. Most such system provide 2 to 3 hours of supply from storage. One has to count well over $1000 per kWh of storage.
- Prices do certainly depend on the installer. The 10 top installers in the United States in 2022 covering some 60% of new installations were the following:
  - Sunrun Inc
  - Titan Solar Power
  - Freedom Forever
  - Tesla Energy
  - SunPower Corp.
  - Momentum Solar
  - Trinity
  - Palmetto
  - Bright Planet Solar
  - Freedom Solar

  Sunrun installations with storage came $5/W or so. Only 5% of its customers got storage. But with 32,000 PV+storage systems it even doubled from the year before.

  Its competitor SunPower offers PV with storage at less than $5/W. Tesla announced in 2021 that it no more sell PV without storage. No wonder, rich of its experience in the electric car business and its leadership in the field with its Powerwall battery.

- Power Purchase Agreement (**PPA**s) became recently very important for the solar PV markets. A PPA is a contract between a corporate buyer (off taker) and the producer (developer, independent power producer, investor) to purchase electricity. The United States had 13.3 GW of signed PPAs in 2021. Four years earlier, there had still not been any. The earlier ones on renewable energies signed since 2008 concerned primarily electricity generation from wind power that was in the early years leading deployment of the renewables.

- The developer of a solar PPA sells the electricity generated to the host customer at a fixed rate that is lower than the local utility's retail rate. PPAs may range from 10 to 15 years and the developer remains responsible for O&M. Following SEIA top corporate solar users were Apple, Amazon, Walmart, Google.

And there is a lot more of solar power to come. The Energy Administration of the United States has issued in September 2021 a new scenario to get the country's electricity supply totally clean already by 2035. A major part of the supply could then come from solar PV.

The new proposal goes back to a study by **NREL: 100% Clean Electricity by 2035** (lead author P. Denhelm) for the Department of Energy. It came with another study of the Solar Energy Technologies Office of the DOE: The Solar Futures Study, with:

- 40% total US supply by solar PV in 2035.
- i.e. 1000 GW in 2035, against 92 GW_{ac} in total by now.
- 225 GW by 2025.
- 550 GW by 2030.
- 1600 GW by 2050 or 3000 GW with increased electrification.
- To reach these deployments, 30 GW_{ac} must be installed each year between now and 2025. Thereafter it will even be 60 GW_{ac} per year up to 2030—four times its current deployment rate.
- De-carbonising the entire US energy system could require 3200 GW_{ac} of solar PV for electrification of buildings,

transportation and industrial energy produced from clean fuels.

- Other key findings from the Solar Futures Study:

  ○ **Electricity prices do not increase through 2035.** Higher electricity prices are offset by savings from technological improvements and higher demand flexibility.

  ○ **De-carbonisation implies significant acceleration of clean energy deployment, with ½ to 1½ million people in solar jobs by 2035.**

  ○ **Storage, transmission expansion, and flexibility in load and generation** are key to maintaining grid reliability and resilience.

  ○ **Demand to be met grows by some 30% from 2020 to 35%.** There will be more electrification of buildings (heating), vehicles, and industrial processes.

  ○ **Land availability does not constrain solar deployment.** 0.5% of the land area of the contiguous US surface area would suffice for all solar deployment projected for 2050, if it were ground-based. But a major part will be building-integrated PV.

  ○ **The benefits far outweigh additional costs incurred.** Cumulative power system costs for de-carbonisation from 2020 to 2050 are $562 billion higher, including the costs of displacing electricity from fuel combustion. However, avoided climatic damages and improved air quality more than offset the additional costs, resulting in net saving of $1.7 trillion. De-carbonisation with more electrification will reduce grid emissions relative to 2005 level by 95% in 2035 and 100% in 2050.

In the meantime, the US EIA projects for 2023 a new PV installation of 29.1 GW from 18.6 GW new in 2022. Additionally, the EIA projects 9.4 GW of new storage capacity for 2023. PV and batteries are making up 71% of new power; PV alone 54%. 41% of new solar installations come in Texas and California. 30% ITC. Wind power installations come further down in 2023 to only 6 GW.

The next decade is declared a "Solar Decade" in the United States.

Challenges must be addressed so that solar costs and benefits are distributed equitably. Solar deployment can bring jobs, savings on electricity bills, and enhanced energy resilience. Interventions, be it financial, community engagement, siting policy, or regulatory measures, can improve equity in rooftop solar adoption. Additional equity measures can address the distribution of public and private benefits, the distribution of costs, procedural justice.

2.2.2 PV Power in Europe

Since the year 2022, European PV markets knew another explosive growth.

The **European Union**, the union of 27 independent nations, without the UK, Switzerland, Norway, Turkey has installed in 2022 new PV capacity of 41 GW, more than in all the previous years. This brought the total PV power installed to 209 GW, both a new record. It's the first time that total capacity of PV passed the 200 GW mark. Only in 2018 the EU had over 100 GW installed for the first time.

The 10 EU countries with the highest PV installations in 2022 in place were the following (new installations in the year in parenthesis):

1. Germany 68.5 GW (+7.9)
2. Spain 26.4 GW (+7.5)
3. Italy 25.2 GW (+2.6)
4. The NL 18.3 GW (+ 4)
5. France 15.7 (+2.7)
6. Poland 12 GW (+ 4.9)
7. Belgium 9.5 GW (+2.5 GW)
8. Greece 5.7 (+1.4)
9. Portugal 4.3 (+2.5)
10. Denmark 3 GW (+1.5)

In 2023 one counts some 60 GW of new installations bringing the total to some 270 GW.

Half of Europe's PV is large scale and the other half is distributed, in particular residential.

Turnover of PV related business in the European Union was €20 billion in 2020. Employment in PV stood at 165,000 jobs.

In 2021, the EU had generated 157.5 TWh from 167 GW installed, approximately one TWh/year per GW.

As part of the overall demand of 2865 TWh in 2021, the EU generated 5.5% from PV. For Germany, Spain and Holland it was over 10%. In 2022 the PV share of total supply increased even to 7.3%.

By 2035 wind power and PV together may meet 75% of demand, at no extra cost.

The analyst **Ember** in London found that the EU generated in the four summer months, from May to August 2022, 12% of its electricity supply, 100 TWh as it were, from PV, more than in any previous year. Gas imports worth $29 billion were avoided that way.

In 2021, the European country coming on top of **PV capacity per inhabitant** was not Germany; it was Holland with 815 W per inhabitant. Germany came second with 706 W per person, followed by Belgium with 544 W.

A lot of European consumers today have become **auto-consumers** with PV electricity covering part of their demand, obviously not totally. They were many more than just a few years ago, such as in

- Germany 1.6 million residential PV plants
- NL 1.5 million
- Spain 1 million+
- UK 850,000
- Italy 630,000
- Belgium 480,000
- France 207,000

In 2021 the EU installed 122,000 panels, most of them imported. Following **Eurostat**, the EU's Statistical Office, the EU has a €6.2 billion trade deficit in PV.

75% of panels imported into Europe came from China; a few percent came respectively from Malaysia, Japan, South Korea,

Taiwan and the United States. Domestic PV module production was 8.3 GW. **Meyer Burger** just decided to increase the module production capacity it operates in Germany to 3 GW a year.

The EU decided in September 2018 to remove import duties on imported Chinese solar panels. Just at a time when the United States had decided to increase such tariffs. 2018 became the year of a new stage of PV deployment in Europe when the decision on the removal of the import duties coincided with the fall of international prices for panels in response to a reduction of subsidies for PV deployment on the Chinese markets.

China is said to have imported to Europe up to 100 GW of PV cells and module in the single year 2022 at a value of over €20 billion. Certainly, not all of this gigantic import could have been installed, and much went into storage houses, but still we see a new age of solar coming. It was triggered by the war in Europe started in 2022 that demonstrated the high security of supply that home-generated solar electricity provides better than any other energy, and in particular gas when it is imported. And even more so for prosumers, the auto-consumers of PV, that gain better independence from increasing electricity rates on the grids.

Solar electricity generation got over 200 TWh in 2022.

Europe does not install any new fossil fuel power stations. Plants with coal are being phased out in the long run while production from gas and nuclear is declining.

In 2020, the EU covered 37% of its electricity demand from the renewables; from PV it was 5.4%. PV power has the fastest growing share: in 2008 it provided less than 1% of electricity.

Nevertheless, much electricity is still produced from existing fossil fuel plants at 41% of total demand it is leading generation, but decreasing over the last years, just like nuclear whose share of a quarter is decreasing as well. Electricity from hydro decreased considerably as Europe had in 2022 the worst drought in 500 years (not many Europeans did complain actually, for most it was a pleasant summer weather).

Fossil fuel derived electricity even raised the carbon footprint of the power sector in 2022 by 8% over the previous year. And carbon prices are high in 2022, they are up to 85 euros/tCO_2 at Europe's carbon market.

At European gas prices of 0.11 euro/kWh, the price of electricity derived from it comes 0.23 euro/kWh, before the carbon price is added, too. Since 2020 PV and wind power electricity have become cheaper than the gas-powered one. Gas-powered electricity is now up to six times more expensive.

In the summer of 2022, 10 EU countries generated a tenth of their electricity from PV: it was even 23% for NL, 19% for Germany, and 17% for Spain.

100% of new power plant capacity in the EU since 2021 is renewable. Two-thirds of it is solar PV. Compared to the new 41 GW of PV in 2022, wind power growth comes behind at a much lower extent than in previous years. Only 14 GW are estimated for wind power additions that year in the EU, not counting the non-negligible new capacity of some 8 GW outside the EU in the UK, Turkey, Russia, and Norway. Still, EU electricity generation from wind amounted to some 400 TWh, still twice the solar generation that year.

The cost of PV generated electricity in Europe is unbeatably low: in the EU is applied the "**merit order**" for the price of electricity, the highest kWh price is applied to all the cheaper suppliers, too. The war in Ukraine led to a scarcity of gas in the EU and an increase of its price from other sources of supply than Russia. As mentioned above, electricity from gas-powered plants increased a lot, making all others dearer as well. **As a result, PV generated electricity that is the cheapest on the EU common market—cheaper than electricity from wind or from nuclear—followed the price increase and so, became profitable.** Hence, the EU Council has decided in October 2022 to put a price cap of 18 cent/kWh to all electricity suppliers, also those that sell above their production cost incurred. **Electricity from PV gets dearer but remains extremely profitable.**

For a new PV sunroof plant, one has to count between €1500 to €3000 per kW installed including all the soft costs. Modules make only a small fraction of the installed plant cost. Details of overall cost depend very much on the country and local conditions; Belgium was given as an example previously, Germany comes later. Plants' amortisation time depend also on tax rebates, Green Certificates or other local support schemes available. It may come as low as 4 years in Belgium. After the plant is paid off, electricity is supplied for free as there are, as a rule, no O&M costs. And lifetime of PV plants is virtually for ever: I have owned myself a little plant in Brussels for 23 years and a large plant on Tenerife island for 16 years without any sign of degradation.

In recent years, battery storage paired with PV knew an explosive market development as well. The EU had in 2022 140,000 new such systems for an overall capacity of 1 GWh. Virtually all the storage market uses Li batteries. two-thirds of the EU market for residential PV+storage was deployed in Germany, followed by a smaller one in Italy. More details are given in a later chapter on PV in Germany.

PPAs may nowadays supply PV electricity for 5.3 cents/kWh. **PPAs enjoy increasing popularity in Europe, catching up with America.** Just as an example one could mention the French telephone provider Orange that concluded a 15-year PPA with ENGIE to buy all the electricity from 2 PV plants in the High Alps region coming on stream in 2022, a 38 MW plant from the L'Epine solar farm and 13 MW from the Ribeyre farm.

Another one was concluded in September 2022. It is a 20-years PPA on some 8 MW between Iberdrola from Spain and Solvay from Belgium. The plant will deploy 100,000 PV modules on 77 ha of rehabilitated "brownfield" land in Saint Fons, Eastern France and will go into operation in 2025, it is planned.

In Europe, solar PPAs for 11 GW in total were concluded in 2021. Spain was again its largest market with 34 projects and 4 GW committed in total.

Eurelectric, the backbone Association of Europe's power sector employing 1 million people, came forward with an interesting projection of how the European power sector is going to develop in the next few years: The European grids will by 2025 have less fossil fuel and less nuclear capacity. New capacity will be primarily PV followed by wind power. From 950 GW by now the EU's power demand will grow to 1550 GW in 2030. From 2021 to 2030 the EU's power capacity has to increase by 62%. By 2030 the EU should add 733 GW of renewable power; a total capacity of 530 GW of PV and 469 GW of wind power should be installed by then.

The objectives had been further boosted by **REPowerEU,** the new Commission's initiative-see below. €84 billion per year must be invested by then. And 23% more in distributed grids.

Progress in de-carbonisation plans: In 2030 the EU's power sector will generate an emission of only 77 g GHG/kWh against 241 g in 2021.

The future of PV in Europe looks indeed brighter than it is already today. The EU is busy taking new groundbreaking decisions under the leadership of the Commission in Brussels.

- **First came the "European Green Deal".** It was decided by the Commission in July 2021. It requires the EU to reduce GHG emissions by at least 55% in 2030. This implies inter alia to increase the share of the renewables: By 2030 the EU must increase, as a binding target, the renewables' share to 40% of demand. 1/3 of the effort needed to accomplish the Deal is supposed to come from the 1.8 trillion € budget of a REPowerEU plan and from the 7-years regular budget of the EU.

- In June 2022 the Commission came forward with this REPowerEU plan: Under the motto: **Substituting fossil fuels and accelerating Europe's Clean Energy Transition.** For a massive speed up and scale up in renewable energy in power generation and other markets, the Commission proposes:

  ○ Increase the target in the Renewable Energy Directive to 45% by 2030 up from the 40% proposed previously. To achieve in the EU a 1067 GW renewable power capacity by 2030.

  ○ Solar PV is one of the fastest technologies to roll out and over 320 GW of it to be installed by 2025, more than twice the rate of 2021. By 2030 almost 600 GW of PV should be installed.

  ○ As part of its increased ambition the Commission proposes also in June 2022 the landmark EU solar energy strategy, COM (2022) 221 (18.05.2022), to deploy 600 GW_{ac} of PV, or 750 GW_{dc}. On average 45 GW of PV must be installed per year. It is the first ever that the Commission has adopted specifically on PV! The Communication foreseen in particular:

- A European Solar Rooftops Initiative. In particular make the installation of rooftop PV compulsory for all new residential buildings
- A PVGIS, a tool for citizens to evaluate their roof's PV potential
- Making permitting procedures shorter and simpler
- Ensuring the availability of a skilled workforce for producing and deploying PV across the EU
- Launching a European Solar PV Industry Alliance

- At last, in November 2022, the EU Commission came up with the proposal for a Council Regulation to accelerate the deployment of renewable energy COM (2020)591 9.11.2022). It finds that solar energy is a key source of renewable energy to put an end to the EU's dependence on Russian fossil fuels while achieving the transition towards a climate-neutral economy. It calls for immediate action to ensure significantly faster permit-granting procedures. The proposal introduces a maximum deadline of one month for the permit-granting process for the installation for solar energy equipment and its related storage and grid connection. It introduces a specific derogation for the systems from the need to carry out environmental assessment under Directive 2011/92/EU.

On October 18, 2022, the Commission published the **7th report on the state of the energy union.** It is the 3rd report since the adoption of the European Green Deal and the first after the adoption of the REPowerEU plan. The report takes note that 2022 is a record year for European PV. And demands the share of the renewables' contribution to overall electricity supply to grow from 37% in 2021 to 69% in 2030.

2.2.2.1 Largest PV markets in the EU member countries

1. Germany

Following the **Photovoltaic Barometer** of EurObservER in Paris, Germany, Europe's largest economy had in 2021 also the largest European solar PV park installed with 58.7 GW, all of them registered on the national grid.

In 2022, latest figures suggest that some 8 GW of new PV brought the total to approximately 68.5 GW, and a deployment of some 350,000 PV new plants. This was bringing the total to 2.2 million plants installed. The state of Bavaria is Germany' top region for PV deployment.

In 2023, the German market increases further by some 10 GW/y and reaches a total of 78 GW.

PV market deployment in Germany is in 2022 3 times higher than that for new wind power, in 2023 it is even 4 times higher.

At an electricity generation of 60 TWh PV provided 12.5% of the country's gross electricity consumption in 2022, 16% more than the year before. Per inhabitant Germany had on average 1400 W/head. Presently PV implies 58,000 jobs. PV revenue of German PV was €1.7 billion in 2022 with over 34.4 mt of GHG avoided.

Germany promotes rooftop PV. By mid-2022, Germany's largest cities had increasing share of PV on new constructions: Cologne or Nuremberg had each some 65% of new buildings with PV, in Berlin it was almost 30%, and in Munich and Hamburg each 10%

For 2030, the government sees a deployment of 200 GW of PV in total.

Germany is a European lead country for PV paired with storage, too. Today over 60% of new residential PV comes with storage. Following **EUPD Research** from Bonn, Germany added 200,000 new such systems in 2022 achieving a total of 667,000. By 2024, 1 million PV+battery systems are projected.

PV is strongly promoted in Germany. The government bank KfW provides with "Erneuerbare Energien 270" favourable credits. Such credits may also include the "soft" costs of installation. Income from PV operation is exempt from income tax. The VAT of 19% is waved for PV plants up to 30 kW.

Since July 2022 the "EEG" feed-in tariff stands at 8.2 cents/kWh for systems up to 10 kW and 5.8 cents/kW for those up to 1 MW. If a 10 kW plant feeds all electricity generated into the grid the tariff paid is 13 cent/kWh for plants up to 10 kW of power.

The kWh price is unusually high due to the merit order effect discussed previously. It follows an EU price cap fixed at 18 cents/kWh.

2. Italy

By year end 2022 Italy had some 25 GW of PV installed, over 2.5 GW came additionally in the year. It's a record deployment for the last few years, actually 150% more than the year before. Nevertheless, some experts complain that in 2011, a year before the European disaster breakdown of the PV markets, Italy had installed 9.4 GW in a single year.

Most PV was recently deployed in Lombardia, Lazio, the province of Lecce, and the island of Sardinia. Viterbo, Cagliari, and Brecia come on top for the PV in cities. Some remarkable large office skyscrapers with building-integrated PV can now be admired in Milan.

More than 28 TWh of PV electricity was generated in the year. 1.12 TWh per GW is, thanks to a better solar climate a lot better than in the more Northern countries like Germany. With over 8% of the total electricity supply in Italy, PV contributed as much as hydro to the supply and 50% more than electricity from wind power. All renewable power together came to one-third of demand.

Italy has currently over 1.2 million PV plants installed. 35% are ground mounted on some 18,000 ha of land. 65% are rooftop. Half of all the PV plants are deployed in the industrial sector, 20% in the tertiary, 18% in the residential, and 11% in the agricultural sector.

Italy is also a major player in Europe for PV+ storage. In 2022 the country had 122,000 PV systems paired with storage, all lithium batteries.

PV is supported by a 50% eco bonus, turned in 2020 into a 110% super-bonus tax credit that is inter alia applied for the PV installation in buildings. Maximum costs of 2400 euros/kW are accepted for support and 1000 euros/kWh of battery. Italy's PV owners are great users of auto-consumption. It is promoted even more so since 2022, for medium-sized enterprises up to 200 kW with a grant of 267 million € and by net metering called "Scambio sul posto".

The feed-in tariff (FIT) has survived from the earlier years in **Conto Energia**. It is also applied to the large ground-mounted projects. But as a result of the electricity price increases in 2022, in Italy, too, PV electricity prices increased as well together with incredible profits, as explained in previous chapters. In reaction to these super profits the government did reduce the FIT paid in 2022, and only for 2022, for plants over 20 kW in size.

At last, one should also mention in PV support a decision by the government to provide a new aid to agrivoltaics with the installation of 375 MW at an expense of €1.5 billion in Italian agriculture, livestock and agro-industry. To the purpose, the finance is provided by the EU Commission.

For domestic production of PV modules, ENEL Green Power plans to build a 3 GW plant at Catania in Sicily. The mega project TANGO at an investment of €600 million.

For the future, by 2030, REPowerEU, described previously, sees for Italy 85 GW of PV, standing for 84% of the renewables in demand, at a revenue of €345 billion and 470,000 jobs.

To conclude on the power generation in Italy, it is worth mentioning that the country has no nuclear power online. And it never had. That makes a big difference with France, the next country on our list.

3. France

France disposes of an operating nuclear park of almost 56 GW. In normal times it generates over two-thirds of the country's electricity needs, and it is the pride of the country—but it is a great barrier to the deployment of PV and all the renewables.

Nuclear power looks indeed very good in view of today's criteria for a desirable type of production: nuclear power is "sustainable" as it emits no GHG. It provides security of supply as all nuclear plants are built and operated in the country. And currently electricity in France comes relatively cheap; clients pay only half for the kWh than the Germans for instance.

Hence, most politicians and most French people alike favour nuclear.

Despite the fact that the nuclear fuel has to be imported, partly from insecure countries, and that production is not particularly clean. The country's nuclear plants are getting old and subject to more repair work: for the winter 2022/23 half of the plants are out of order. For the total year 2022, the losses of the operator of France's nuclear plants EDF amounted to €17.9 billion. And a mountain of expenses is in the bush for eventually dismantling all those plants and storing safely all the radioactive residues.

And keep fingers crossed that there will not be an accident despite tough security measures.

The French President and a majority of politicians are in favour of launching the construction of a new park of the next generation, the EPR reactor plants. Leaving alone the fact that its prototype in **Flamanville** had its construction time delayed by ages and its budget boosted by billions of €, It will have cost a minimum of €13.2 billion and will not be grid connected before mid-2024, a disaster.

Still, PV has a long tradition in France. The country was once a European pioneer for its promotion and the deployment of the first markets. And until today a powerful community of supporters makes their voices heard.

By year end 2022, France had 15.7 GW of PV installed, of it 14.8 GW in mainland France. It added 2.7 GW to its park in the year. France had 600,000 PV plants registered at the end of September 2022. Of the total, 208,000 were auto-consumers, twice as much as 18 months ago. In 2015, it had only 300 of them.

The Nouvelle Aquitaine, in the South, came on top of the installation rate. For the Paris area a solar cadastre giving all local solar radiation values is available.

In the first half year of 2022, the country generated 9.6 TWh of electricity: it was 3.8% of demand. In the best month of June, it was 6%.

France has a Feed-in Tariff for PV. The tariff varies slightly every 3 months. When, for instance, a plant sold all the electricity it produced to the grid, later in 2022, it was 17 cents/kWh generated for a power of less than 9 kW, and it was 11 cents/kWh for a power less than 500 kW.

For auto-consumption, the electricity consumed is not taxed. The surplus sold to the grid is paid 10 cents/kWh. Smaller systems receive in addition a premium of €320 for less than 9 kW, and less when the system is larger. The income from the sales to the grid is fixed for 20 years by contract, part of it is subject to tax. Most French residential PV plants employ auto-consumption with re-injection of the surplus energy.

PPAs in France were already mentioned in a previous chapter. In November 2022 at last was signed a very big one between Voltalia and the carmaker Renault. It is a 15-year PPA of 350 MW of PV to be installed until 2027.

Future outlook: France foresees 33% of renewable energy contribution to cover its demand by 2030. PV is supposed to dominate new installations. **18 to 20 GW of PV by 2023 are projected.** But the "acceleration" law for the renewables of January 2023 decided by the Parliament is rather a deceleration law.

In the country's multi-year energy program PPE are foreseen a thousand PV projects on public land by 2025. Some along motorways, others on parking shades.

By 2050 President Macron wants 100 GW of PV in France.

4. The Netherlands (NL)

NL has only one little nuclear power on its grid, just one plant, and a small one. National energy policies in Europe between countries are so different from each other that the EU long hesitated to come along with a common energy policy.

NL is keen to cut GHG emissions by 40% in 2030. The last coal power plant is to be closed a year before.

The country's ambition is to have by 2023 a share of 16% sustainable energy supply.

NL passed in 2022 France as the 3rd largest PV market of the European Union. At year end 2022 it had 18.3 GW installed, + 4 GW in the year—compared to 15.7 GW cumulated in France. The strength of the Dutch PV market is surprising for a country so much smaller than France. But it is not surprisingly that the country comes on top of PV capacity installed per head in the EU: with 825 W/head it comes even on top in the world just behind Australia. Most of the cumulated PV installed at 18.3 GW comes in the power range above 15 kW. Still, 1.6 million Dutch houses for a total of 5.6 GW are equipped with PV. A million power customers are engaged in energy co-operations.

PV generated electricity provided in 2022 14% of demand. In August that year it was even a quarter of demand.

The country employs several instruments for support of PV, net metering, tax benefits, and investment subsidy:

- Net metering of PV electricity is a great stimulus. At current high electricity prices on the grid auto-consumption with sale of the surplus is highly profitable. It is the reason that the country did not develop its PV systems paired with a battery.
- The VAT of 21% for residential PV can be reclaimed on demand.
- Investment subsidy is granted in the frame of SDE++, a government program for stimulating sustainable energy production with a total of €13 billion in 2022. In this round, PV projects of 2.26 GW in total were selected in the bidding process. The grants are given for the feed-in of 50% of peak capacity. 1.3 GW of the total for rooftop application and some 950 MW for ground-mounted ones. 20 MW came in support of floating PV.
- An example is a PPA getting a support for a dairy co-operative and an energy company installing 294 MW on the farms.

The potential of PV in the future was analysed in a report published by the Dutch **TNO** in April 2022. In two scenarios TNO sees for the country by 2050 a PV market of 55 GW, resp. 132 GW. Demand will increase from 110 TWh by now to 500 TWh. It will not lead to higher electricity cost, they found.

5. Spain

The country came in third position of the EU following the volume of its PV market in 2021. But Spain benefitted from an explosive growth of PV deployment starting in 2022. The country had already such a market rush of PV more than a dozen years ago; this time it looks more sustainable. At year end 2022 Spain took with **26.4 GW** installed and an additional 7.5 GW this year the second place from Italy. Spain is indeed well suited for solar deployment: its solar climate is the best on the continent, its landscape and the concentration of population are favourable. Large parts of Spain being arid and less developed Spain could eventually become "**Solar Arabia**" in the sustainable solar world of tomorrow.

Following the grid operator REE, Spain had in September 2022 a total capacity of 17.2 GW of large-scale PV installed, to which comes 2.5 GW of PV in auto-consumption. The latter sector was recently particularly prolific. In 2022 PV generated 12% of demand, from May to August it was even 17%, the EU's second highest after the NL.

Spain supported its renewable energies starting with the "Climate Change and Energy Transition Bill". A new support for PV came in 2018 when the government abolished the "sunshine tax". That one had obliged the auto-consumers to give their surplus electricity back to the grid free of charge. A new support mechanism was also introduced for dispatching PV on the grid.

Auto-consumption through rooftop PV and energy co-operatives are now booming in Spain. A solar home system can currently save approximately 50% on an annual bill. In 2021 the government provided €1.2 billion of subsidy, too, inter alia for auto-consumption plants.

There are also fiscal aides, for instance, personal income tax credit for installing PV for auto-consumption. Spain has no net metering: electricity injected into the grid is paid by a lower tariff by REE.

Spain is a big promoter of PPAs for PV. The biggest one in Europe is the Francisco Pizaro solar farm near Caceres in Estremadura built by Iberdrola. With 590 MW it is bigger than another mega PV plant in Estremadura, the Nunes de Balboa with 500 MW. The new one will deploy 1.5 million modules on over 13,000 trackers feeding into 313 inverters, the total on 1300 ha of land. The cost is €300 million, well below 1 dollar/W. Iberdrola works with an ornithologic society to identify and protect bird breeding sites. Archaeological remains are identified and will be protected. The energy generated will be supplied to Danone, Bayer and Pepsicola, the partners in the PPA.

Energy co-operatives are helping communities to overcome the initial expenses. In Zaragoza, the capital of Aragon, the Ecodes co-operative works with utility EDP and the local government to launch a "barrio solar", a solar neighbourhood. PV will be installed on the municipal buildings generating electricity for hundreds of homes. Participants are free from the initial investments but pay a monthly quota of €6.90. They pay then their electricity 30% cheaper than the market rate. To help the most vulnerable, the municipality has chosen some families that get the electricity for free.

For the future, Spain has a target of 72 GW of PV by 2030. It comes as part of 74% of renewable electricity by that time and 100% in 2050.

6. Poland

Here comes again another type of EU energy country: Poland is not a nuclear country, it is a coal country, the EU's biggest coal user. A third of the EU's electricity from coal is produced in Poland. Poland comes among the most polluting nations. Coal plants are old and subsidised with billions of euros. The World Health Organization (WHO) found for 2017 that Poland has 36 of Europe's most polluted cities. It all started in communist time.

But still, there is some hope: In recent years, the consumption of coal was decreasing. Poland's electricity supply in particular is getting "greener": in 2021, only 72% of electricity demand was met from coal plants; and so far, no new ones are being built. The decrease of coal power is encouraged by high CO_2 prices of some \$63/tCO_2. Renewable electricity came in for 17% of demand already, the rest being natural gas.

PV power and PV generated electricity are currently the fastest growing in Poland's energy world. And the growth is very recent: only in 2019 the country deployed the first 1 GW of PV capacity. Poland did not participate in Europe's first big boom of the PV markets at the turn of the century. Interest started slowly from 2004 on when the country had joined the European Union. *I remember, when I first visited Warsaw in the early days, they had only one PV panel installed, at the University.*

It's a different story today. PV has become popular in Poland. The country celebrates "a golden decade of solar power". Since 1 GW in 2019, PV markets in Poland have exploded. 3.9 GW only a year later, then in 2021 PV at 7.7 GW passed Polish wind power. The rush continued with 12 GW in 2022.

Poland's PV market growth by new GWs of power installed is the fastest in the EU behind Germany and Spain.

Most spectacular is the growth of auto-consumers; there are over 1.2 million of them in 2022.

First, Poland's PV markets were promoted by a **rooftop incentive programme**. So far 410,000 domestic PV systems were supported. In the latest round for applications from December 2022 the subsidy for PV systems and storage is considerably increased. Until April 2022 prosumers benefitted from net metering, later transformed into "net billing"—the latter is paid for the market value of the PV electricity and not its volume. PV benefits from 8% VAT instead of 23%.

In 2022, with the war in Ukraine and restrictions in the gas supply, profitability of PV increased further, like in the rest of the EU, in a market of high electricity prices.

Further support comes from "CfD" contracts by the government after tendering for systems respectively under 1 MW, and above 1 MW.

Since 2021 PPAs are booming in Poland as well. Examples are, e.g.

- A PPA between Axpo Polska and R. Power Group with Nomad for O&M. Various PV plants of 300 kW to 30 MW, in total for 300 MW
- 65 MW for a 10-year PPA between the Germans Baywa re. and Heidelbergcement at Witnia near Posnan
- A 10-year PPA of 36 MW between the Danish Better Energy and the Norwegian Statkraft in Resko, without subsidy as the partners underline, and 17 kt of GHG emissions avoided per year

For the future, Poland's energy situation remains unclear. There is the EU constraint of net zero GHG emissions by 2050, but... For PV no realistic target has been announced. The national grid is not in very good shape to accommodate much grid—connected micro-PV. The current government is planning for new coal power plants and even nuclear plants are on the drawing board for Poland. Future will show.

2.2.2.2 The largest European PV markets outside the EU

1. United Kingdom

Now after it has left the EU with the Brexit, Britain does well continue to be a key player of PV deployment in Europe. The UK had at year end 2022 in total 15 GW of PV installed, just a GW less than sunny France at that time.

Like everywhere else in Europe, the year 2022 has brought explosive growth to the British PV markets. As a result of big inflation, supply restrictions of natural gas and the following increase of electricity prices, solar PV generated electricity as a homegrown asset available even for self-consumption, has become most attractive and profitable. And as described previously, PV electricity has become the cheapest on the markets. As a result, PV power is being deployed at new strength, also in a rainier country like England and the whole UK. And a lot of it is coming unsubsidised by now.

While PV markets explode, coal markets are shrinking in Britain: in the summer of 2022 the country generated for the first time ever more electricity from PV than it did from coal!

PV capacity deployment accelerated in 2022 by some 1.5 GW. Almost all of new power capacity installed in the UK in 2022 was PV.

The UK had then over 1.4 million certified PV installations, over 1 million of them residential. 3.3% of the 29 million British homes have PV.

A third of capacity is rooftop. Half of the rooftops come as residential, the other half is commercial and industrial.

Households can save in the range of £1000 per year as auto-consumers, it is said. The value of a house may increase as well.

In 2019 the FIT ended in Britain and the PV market was knocked down at that time. Net metering came in a new form later: in 2022 the replacement is called Smart Export Guarantee (**SEG**). It provides payment to the excess electricity from prosumers to the grid at a special tariff.

Since 2022 VAT has become 0 for PV.

Approximately two-thirds of installed PV capacity is ground mounted. Large ones can be supported by the CfD scheme (Contracts for Difference). The UK has over 500 solar farms. There are many MW-size PV plants, the 20 largest in 2022 have an overall capacity of 876 MW. The greatest one is the **75 MW Llanwern Solar Farm in Wales**. The German PVDP is proposing an 840 MW PV plant in Oxfordshire for 2025. An even larger of over 1 GW capacity may be built in Romania by the UK developer Rezolv Energy.

Cornwall is the best area for solar energy in England. It has over 8000 solar sites.

The average electricity generation on 1 kW PV over the year in the country's solar climate is 700 to 900 kWh. It is similar to that of Germany. The PV capacity or load factor in Britain is 10% (for offshore wind it is much more favourable at 37%, one of the best in Europe). Later more on the role of capacity factors.

Generation of electricity from PV plants is booming together with the capacity extensions.

For the future, by 2030, 32% renewable electricity are projected in Britain like in the EU Directive on the subject. By that time 80 to 120 GW of PV should be installed.

Eventually, one could generate all electricity from PV: 100% would in principle be possible from PV on 12% of the British landmass. But nobody wants a PV monoculture for electricity supply...

2. Turkey

Turkey has by now the smallest PV market of all the European countries that we reviewed thus far. And by contrast it has the strongest solar resource: the many Sun-seeking tourist in Turkey can tell.

Still, Turkey had a nice PV park of 7.8 GW in operation at year end 2021. And it generated 13 TWh from these plants in the year, at 1.7 TWh per GW installed a record in Europe.

PV makes currently 8% of the country's overall power capacity.

So far emphasis was laid on large plants in the MW range. A 140 MW PV plant on a steel mill is the world's largest. Its purpose is the production of green steel. The plant to be completed by April 2023 will come on a 630 m^2 roof. It is built on a plant of the company Tosyali in partnership with Solar Apex and the Chinese Huawei.

Tendering of large PV plants resulted in some of the world's lowest PV electricity cost at 3 cents/kWh. That one involves a PPA for 15 years.

More recently Turkey put more emphasis on residential PV applications. Since 2020 auto-consumption is promoted by net metering. In 2021 there was applied a FIT of some 3.7 cents/kWh. Another particularity of Turkish PV that differentiates it from other European countries is its recent involvement in the production of cells and modules in large GW-size factories. The latest is a 2 GW plant in Izmir applying modern silicon technologies such as PERC cells and "half-cell" modules.

For the future, PV capacity in Turkey, following its Solar Energy Roadmap, may reach 38 GW by 2030.

But the terrible earthquake in February 2023 with some 50,000 deaths is expected to bring all PV deployment to a halt. Turkey has to care for its survival needs. Perhaps some of the new buildings can get PV roofs?

In 2019, the author published together with his friend Prof. Tanay Sidki Uyar the book The Triumph of the Sun in 2000–2020 *in Turkish.*

3. Switzerland

The author at a speech in Switzerland.

The PV boom around the world goes on. This time in an alpine country: By 2022 Switzerland deploys almost 1 GW of additional PV, a 50% increase over the PV market supplement in the previous year. This brings the total PV capacity installed to over 4.5 GW. 517 W per head, a value that puts the country high up in Europe, only a bit behind The Netherlands that come currently on top.

15% of all the power capacity installed in the country is now PV. There are 150,000 PV systems in operation. Every third of them was paired with a battery: at year end 2021 Switzerland had in total a PV storage capacity of 157 MWh.

For auto-consumption the FIT ranges currently at 10 cent/kWh. In the absence of net metering, prosumers have to install a second meter for the electricity they buy. In Switzerland electricity prices are exploding like everywhere else in Europe and a PV rooftops gets all the more profitable.

The largest PV plant came into operation late in 2022. It has a capacity of 7.7 MW feeding into a refinery. The ground-mounted modules are Swiss made by **CSEM**.

Perhaps because of shadowing effects in this mountainous country generation of PV electricity is more on the modest side with some mean 0.8 TWh per year for a GW of PV installed.

For the future, all the country sees a further booming development for PV.

By 2050 PV could cover 50% of the total electricity demand, finds the association Swissolar. It would be enough to cover 40% of all roofs in the country with PV for that.

2.2.3 PV Power in China

China suffers particularly from climate change. Between June and August 2022, the whole country was affected by the most severe heat wave experienced anywhere in the world. Besides the drought that came with it, there was also unusual heavy rainfall in the North of China.

China is committed to combat climate change and limit GHG emissions: the President has announced in 2020 carbon peaking by 2030 and neutrality by 2060. Part of the policy is Beijing's commitment to deploy renewable energy supplies, and China became the world's giant in PV production, domestic installation and export, no matter any COVID-19 restrictions:

Wonderful China: The Li river.

China deployed more renewable power capacity in 2021 than the rest of the world combined. In the first 6 months of 2022 PV accounted with 31 GW for 45% of new power capacity in China, the highest of all new power additions, together with wind power, coming second behind, and hydro even 77%. PV deployment increased by 137% this year compared to the previous year.

In September 2022, cumulative PV installed topped the 350 GW mark, overtaking for the first time wind power capacity by a small margin, as shown for all capacities by final 2022 in the following:

- Coal 1311 GW +45 GW
- Hydro 403 GW +14 GW
- **Solar PV 392 GW +87 GW**
- Wind power 344 GW +37 GW

PV plants in China.

PV provides already over 15% of power capacity in China against 0 it was 20 years ago. Following **Apricum**, an analyst of the renewables in China, solar PV will become during 2023 power No 2 in China before hydro and wind power.

According to a statement released jointly by the National Development and Reform Commission (**NDRC**), China's top economic regulator, and the National Energy Administration (**NEA**) at the end of May 2022, China will increase total installed wind and solar capacity to over 1.2 billion kW (**1.2 TW** or 1200 GW) by **2030**. Already by 2025, the country will generate 3.3 trillion kWh (3300 TWh) from the renewable energies to further boost its green energy transition within the 14th Five-Year Plan period. The renewables will further replace part of power capacity from fossil fuels by 2030.

In the year 2022, a record PV capacity of some 87 GW was newly deployed in the country. Almost two-thirds of it distributed, building integrated, commercial & industrial and residential; only a smaller share for large-scale utility-scale PV plants. This is a happy trend as previously ground-mounted PV was the priority. In the past this was leading to abuses in the Northern areas of the country where the whole of large mountain regions were covered with panels with the risk of degrading the landscape.

The electricity generated from the PV capacity installed by now comes at a modest 1 TWh/year per GW installed, the solar resource and accordingly the productivity near the big demand centres in Eastern China being not really better than in Western Europe such as Germany.

Analysts said that China's plan to further optimise its energy mix by building large solar and wind power facilities in the country's desert areas will facilitate the ambition to reach the 1.2 TW goal by 2030. The **Western provinces** like Sichuan, Yunnan, Gansu, Inner Mongolia, or Tibet are better suited for the deployment of large PV plant: they have an ample solar resource on huge areas of land, some being deserts or barren land. The central government has taken a proactive stance and initiated, in the frame of its 14th Five-Year Plan since 2021 something new and very big, the "clean energy Bases" to be deployed on such type of land.

The first "100 GW Base project" will spread over 19 provinces, all commissioned by year end 2023. Massive projects include next to PV also wind power and an option for green hydrogen generation.

A second batch of 455 GW in the Gobi desert and similar places was approved in spring 2022, to be ready by 2030. The investment cost is estimated at $450 billion (3 trillion yuan).

A 3rd batch is already being discussed.

It is also planned to build **ultra-high voltage transmission lines** to connect the new renewable power centres in the West of the country to the demand centres near the East coast. They will likely hit 300 GW of transmission capacity by 2025 up from some 200 GW in 2021.

PV will also have its role to play in areas envisaged for industrial transformation, like the coal country of the Yellow River Basin that covers Gansu, Henan, Inner Mongolia, Ningxia, Qinghai, and Shanxi province. In this large area investment for clean energies like PV is also supported by fiscal measures of the government.

As mentioned before, things have changed since a year or two, and distributed has by now become the top runner of PV deployment in China. Already in 2021 much of the national PV markets were on rooftops. For domestic demand of PV central and provincial governments, too, have been proactive since the onset of the 14th Five-Year Plan in 2021. That year NEA initiated a new pilot program specifically on distributed PV: By the end of 2023 there is a revolutionary storm of rooftop PV coming. By then already 50% of all state-owned buildings should have PV panels integrated, 40% of the public buildings, 30% the service centres and 20% of houses in rural areas, in 676 counties. By 2030 half of all Chinese buildings are supposed to have PV power panels integrated.

In complementation to NEA's initiative, 29 provinces have released official plans of PV implementation: by 2025 they are aiming at an installation of 390 GW of power in total.

Many **local governments** such as Zhejiang or Chongqing, have taken such initiatives. Local authorities partner with developers, often state-owned companies to meet PV rooftop targets. Either developers lease rooftop space to install PV, or owners purchase the PV panels and sell (part) of the electricity they generate to developers. Local jurisdictions may offer one-time PV procurement subsidies or a feed-in tariff, the latter for 2 to 3 years.

Provinces have also initiated new solar PV industry development plans. The Province Guangdong has released a first "Silicon Energy Development Action Plan 2022–2025".

Board meeting in Zhangjiakou, 2019: First row, the author in the middle; on the left of the row, the author's friend Qin Haiyan.

The State Council of China has designated Zhangjiakou, a site of the Olympic winter games in 2022, to become the country's first National Renewable Demonstration Zone. It is to provide an opportunity for power sector reforms in favour of accelerating the scale up renewable electricity generation and use.

Several industrial parks address C&I integration often aiming at 80% coverage by 2024.

In support of all the ambitious development plans there is also some fiscal aid provided. As an example, in August 2022 the State Tax Administration issued a guideline for "preferential tax and fees policies to support green development". Some of them address solar PV.

Then the Ministry of Finance (MOF) released a notice on the "fiscal and taxation support plan for promoting ecological protection and high-quality development".

China's PV manufacturing industry has developed at breathtaking speed. Following NEA $22 billion were invested just in the first 10 months of 2022 in the PV cell and module industry.

2.2.4 PV Power in Japan

In 2022, at some 85 GW of capacity installed, Japan is the world's 4th largest PV market after China, the EU, and the United States.

Among the country's different sources of power, PV's capacity comes second after LNG, but even before coal, and in decreasing order hydro, nuclear, oil, wind, and bio-power. 20% of PV capacity installed is residential.

In terms of electricity generation, PV comes after LNG and coal power, but still before hydro, nuclear, bio- and wind power.

Since 2014, renewable electricity's share to total generation doubled to 22.4%. Most of the increase was due to PV: virtually all of new power capacity installed in Japan in the last 6 years was PV.

PV's electricity generation is highest in May, in the summer come the monsoons. All Japanese regions are favourable for PV, and in particular Hokkaido...

At the time when the **Fukushima accident** happened in 2011 and nuclear generation disappeared, PV started its rise when the FIT was introduced. In the meantime, the government was busy reviving nuclear power again. From 57 nuclear plants in total, 24 will be definitely decommissioned, but 10 are back in operation and 3 new ones in construction.

In April 2022, Japan changed its PV support from the FIT to the feed-in-premium (FIP). With the ultimate goal of achieving market parity of PV, the government introduced a **"Feed-in-Premium" a FIP**, nothing else than a fixed bonus to selling to the grid or via a PPA. Solar systems larger than a MW fall under the FIP program, for smaller ones it's optional to keep the FIT.

While under the FIT scheme utilities are obliged to buy renewable electricity at a fixed price, irrespective of the prevailing market price. Under the FIP, utilities are no longer obliged to purchase the electricity at a fixed price. Generators have to sell at the market price that is fluctuating. A fixed premium is offered to augment profitability. Incidentally the use of storage is encouraged this way: the producer having a battery paired with his plant will rather sell when the market price is high and store his electricity when it is low.

A PPA market for PV exists as well, be it for schools, banks, or shopping complexes.

It is important to note that over 50 companies offer by now the installation of PV without initial cost.

For the future Japan keeps its objective to make renewable electricity—with PV on top—the main source in the country:

Since October 2021 the government has adopted a **6th Strategic Energy Plan** to be revised after 3 years. A first objective is to make Japan Carbon neutral by 2050. In an ambitious outlook the share of renewable electricity should double by 2030 to 36, 38% of total supply.

- By 2030 PV should come to some 15% of generation—from 9.3% in 2021.
- Hydro should then reach 11%, wind power and bio-power respectively 5%.
- Nuclear power should then be at 20 to 22% of supply. Electricity from the fossil fuels should decrease.

Following RTS Corp, METI has in 2022 given further support to make PV the maximum part of domestic generation and 120 GW_{ac} of additional PV capacity installed from now until 2030.

To close with a personal note, I went in the past frequently to Japan in particular to attend PV meetings as an EU Official in different parts of the country. For installing a PV generator on my house in Brussels in the year 2000, I bought there and brought the solar cells, some of them were golden in colour, others blue, from Japan and got them assembled to modules at home.

2.2.5 PV Power in India

Needless to remind that India has a warm climate, most of the winter months, too, as it is blessed with a lot of sunshine. But for its power supply, the subcontinent relies mostly on coal. Over 50% of India's electricity is produced from coal. The country seems not to be keen to give up on it very soon: unlike most other big countries India wants to become carbon neutral only by 2070.

But by now, India has taken the route of a gigantic solar revolution: from a tiny 11 MW PV market in 2010, by now, just some 12 years later, the country has achieved a nationwide installation of 62,000 MW, a thousand times more of it, or 62 GW. Coal comes still on top of India's national power capacity, but at 15% solar PV beats by now, and every year more so, all the others: it is the highest among all the renewable capacities with hydro-, wind-, and bio-power, that have reached a respectable 40% of India's total power capacity. PV power comes currently even 10 times the country's nuclear capacity.

In December 2022, India's Ministry of New and Renewable Energies announced that the country has by now a total of 62 GW of PV power installed. The Ministry noticed that solar tariffs have reached grid-parity in India.

52 GW of PV are ground mounted, 7.9 GW are on rooftops, and 2.1 GW are off-grid. For the latter are summarised the following applications:

- For 216 MW stand-alone PV plants
- 287,000 solar water pumps
- 8 million solar lanterns
- 1.7 million home lights

In 2022 total PV power deployment increased by over 13 GW, more than any other source of power did. In the financial year 21/22, $14 billion was invested in the renewable power.

Much of the solar PV was installed in solar parks in the dry and desert-like regions of **Rajasthan, Gujarat, Karnataka**. One of the world's largest PV plants is the Bhadla Solar Park situated in the Jodhpur district in Rajasthan with 2.25 GW. It was built since 2015 at a cost of $1.4 billion.

Two-thirds of India's residential rooftop is installed in 2022 in just one state, Gujarat; half of it subsidised, the other not.

Interesting is the trend for "Supply RTC (Round-The-Clock)" attempting 24-hour electricity supply by combining PV and wind power in hybrid systems on a large scale. In a later chapter we come back to the potential of continuous electricity supply this way. A 30 GW hybrid system of PV with wind power was inaugurated in December 2022 in Gujarat by Prime Minister Modi. 30 GW in one plant, not 30 MW! A world record. It cost over $20 billion. Gujarat is also the Prime Minister's home state.

PV modules are preferentially provided for PV projects in India from domestic producers, such as Waaree, Adani Solar, or Goldi Solar.

Electricity generation from PV in 2022 was 7% of the country's total supply. It grew from April to October 2022 28% over the same period the year before. It was not far from the generation from wind power.

And the future? India approved in September 2022 a new 5-years National Electricity Plan. In 5 years, PV power should increase again by 27%. At the current pace of deployment, it might even be more.

The Renewable Purchase Obligation (RPO) is to be extended until 2030.

2.2.6   PV Power in Australia

Australia is a solar PV country par excellence! At least it has become one just recently. By now the country has 3 million solar rooftop installations 10 times more than 10 years ago. 30% of all Australian houses have PV on top. PV is the fastest growing electricity generation source, in the future there will even be a lot more as it grows massively.

In early 2022 Australia had in total 26 GW of PV installed, 4.9 GW more than the previous year. Rooftop PV came first at

16.5 GW, + 3 GW in the year. PV is the fastest growing power in Australia—like in many other countries as we have seen. Australia's PV has been now reached the power capacity with black coal, the traditional frontrunner. And 9 GW PV power additionally as utility scale with +1.7 GW in the year.

On average a rooftop has 8.8 kW power in Australia. Most of them are residential, a smaller part comes for commercial and industrial purposes. Owners receive a reduction in their bill after feeding electricity in the net.

Nationally rooftop PV capacity is as high as hydro- and wind power combined. Since 2016, Australia installed preferably wind power. Only since 2018, just 4 years ago, solar PV got the priority in installation.

For the time being, black+brown coal come still first at 58.3% in 2022 in the large network, the NEM, that covers all Eastern Australia. In total generation in TWh and contribution to supply in % come as follows:

- Solar PV 29.9 TWh, 14.3%
    - PV rooftop 18.6 TWh, 8.9% (19% higher than in 2021)
    - PV utility scale 11.3 TWh, 5.4%
- Wind power 26 TWh, 12.5%
- Hydro power 16.7 TWh, 8.5%
- Coal, black and brown 124 TWh, 58.3%

Of the total 208 TWh of the NEM the renewables came with 73 TWh, 35%.

For 2030 Australia projects an additional PV capacity of 60 GW bringing the total to 86 GW—at the expense of coal capacity. Renewable generation's share should then double from the 35% now to 70%.

2.2.7 PV Power in Southeast Asia

In South-East Asia in 2021/22 a total of some 60 GW stood installed. Approximately the same market volume as India.

S. Korea and Vietnam are the two countries that suffered from terrible wars not long ago. And in particular these two stand out in S.E. Asia now as those that did take the train of solar installations on a large scale. They moved in very recently— Vietnam had in 2015 just 4 MW installed—and they are bound to go ahead with massive deployments.

S. Korea and Vietnam have respectively by now some 25 GW installed. It is for each of them the same PV market volume as Australia's.

South Korea relies for its electric supply mostly on fossil and nuclear power. PV generates only some 4% of supply today. But following the 9th Basic Plan for Long-term Electricity Demand and Supply the renewable electricity's share of supply should mount to 40% by 2034 from 15% at present.

The country is the home of Hanwha, one of the world's largest providers of PV cells and modules. The company was once inherited from Germany.

PV power capacity came 25 GW including additional 4.4 GW newly installed in 2021.

By 2030 the country plans to install 34 GW more of PV.

Vietnam had by early 2022 some 25 GW of PV installed; 11 GW have been installed alone in 2020, and 6.5 GW added again in 2021/22. Most of it is installed in the South.

PV reached an impressive 25% of the total capacity in the country. Vietnam has already 10,000 solar rooftops; and it had in 2020 already 15 utility-scale PV plants installed. Among the renewables PV is the favourite: wind power in Vietnam came only in at 11.8 GW. But together PV and wind power are planned to provide 50% of supply by 2045.

The rest is hydro, the part of coal that is by now the leading supply will be shrinking. Atomic power is planned but fortunately keeps for the time being on the drawing board.

PV is promoted by the FIT and PPAs.

Following McKinsey the country may have over 60 GW of PV installed by 2030.

The PV markets in the other S.E Asian countries are still in an initial stage.

Thailand has 3.5 GW installed. The country comes second for solar installations among the ASEAN countries. It introduced year end 2022 the FIT that may accelerate deployment. Recently the group Constant proposed three blocks of PV of 180 MW in total with storage and at a FIT.

The Philippines had some 2.5 GW of PV installed by 2022. Large plants of several GW capacity are planned by "**Solar Philippines**".

Malaysia, which is producing PV cells and modules under license had 2.2 GW of PV installed in the country.

A tourist site in Malaysia.

Cambodia has 20% of households still off-grid. A few hundred MW of PV are in operation; PV electricity covers 6% of demand.

Laos is the country of hydropower. The PV market is still in its infancy there.

PV in Indonesia is still in the very beginning. It has not more than 200 MW of PV installed. But the potential of this huge country will a multitude of disconnected islands is promising.

2.2.8   PV Power in Latin America

At year end 2022, Brazil had 24.9 GW of PV installed in total. It is the maximum of PV capacity installed in Latin America at this time.

Yet in 2016 it had only been 150 MW in Brazil.

PV market growth goes ahead at an incredible speed by now: In 2022 almost 9 GW have been added to the country's solar capacity, lately not less than 1 GW additionally per month. Of the total, 17.2 GW stand as distributed PV, half of it for residential consumption; 7.8 GW are utility scale. Several PV plants, in the range of 200 MW, were installed in the States of Bahia and Piaul by ENEL Green Power.

The deployment of PV in Brazil created 630,000 new jobs. It mobilised a budget of $20 billion.

For the country's electricity supply, PV at over 10% of it comes now third behind hydropower that provides the lion's share with over 50% and wind power. But PV exceeds supply from fossil sources already.

On a personal note I may add that I came myself to Brazil some 50 years ago with a French governmental delegation from Paris. I remember proposing a large PV initiative: In vain, it was the time when the country started its Bio-ethanol for transport program, another unique success of Brazil. And now PV: you can count on the Brazilians to get things right.

Mexico comes close to 10 GW of PV in early 2023. Most of it is utility scale. There are 60 large such PV plants. The 750 MW plant at **Villanueva** is even the biggest in Latin America.

In Mexico too, PV is a newcomer despite its ample solar resources. 5 years ago the country had in total only 170 MW of PV installed. By now the target is to double PV and wind power capacity together from 17 GW to up to 40 GW by 2030.

Chile is famous for its Atacama desert, one of the driest places on Earth and for its clear sky used for some important space observatories. It is interesting to note that in this desert Chile installed an enormous **CSP** project, a central solar receiver on a tower with thousands of heavy reflecting mirrors around. But electricity generation at 5 GWh in 2022 is ridiculously small compared to large PV plants.

In 2013, Chile had only 1 MW of PV installed. But since that time only PV and wind power were the new capacities installed in the country. By now PV capacity has reached over 5 GW providing already over 20% of the country's electricity supply. Just to mention the 180 MW_{ac} PV block at **Coya** in Antofagasta, in the northern desert where the French ENGIE has installed the continent's largest storage system of 640 MWh.

For the first time in 2021/22 PV and wind power generation together at 27.5% of the market have overtaken electricity from coal that stood at 26.5 GW.

2.2.9   PV Power in Africa and the Middle East

PV in Africa is not comparable with the rest of the world. First, there are still living hundreds of millions of people without access to electricity and to electric light. Since ages PV was recognised as THE opportunity to change things to the better, to provide potable water with PV water pumps and Solar Home Systems (SHS) for a minimum of comfort. In vain, the little that has been done was a drop on a hot stone.

PV has become a great success story, but unfortunately not in Africa. It would be a misunderstanding to think that all efforts failed because a lack of money; Africa has its own cultural and social environment, a lack of political leadership and foresight. Insecurity and chaos left behind by the colonial powers.

In Africa, the people suffer simply from the feeling being left behind, a lack of progress. A lot of political talking, not much PV installed, just a hand-full of GWs for the whole continent in lack of power. Namibia, Botswana, Kenia, Tanzania, Uganda, you name them, all with an excellent solar resource that remains untapped. Much political planning for the future, many GW of PV on the drawing board, yes, but not much realised.

A bit light at the end of the tunnel comes from South Africa where a number of PV plants of close to a hundred MW in size have been installed and are working.

The Arab countries in Northern Africa and the M. East benefit from a nice solar resource, but first solar projects were CSP ones, for instance, in Egypt, or in Morocco the Noor projects, in total for 500 MW, and even with an encouragement from Spain and Germany to build the first ones. Solar towers with a multitude of mirrors around are impressive to look at, a demonstration of solar enthusiasm. But the countries took some time to realise that CSP is a dead end route to solar electricity, the real thing is PV.

For the time being, PV rooftops are still relatively rare in all the Arab countries. PV is a business, and potential benefits for the people are so far left aside. The preference goes for large utility-scale PV plants, mammoth ones, mostly in the arid land, the deserts and there are plenty of those:

- The largest PV block is being built in the desert of the **Emirate of Dubai, the Mohammed bin Rashid Al Maktoum Solar Park.** It has already 1.6 GW operational. The final capacity will be 5 GW by 2030.

- In **Egypt** is promoted since several years the Benban PV project with 1.8 GW of capacity as the ultimate target. It is situated in the Aswan desert, with access to the Nile. A small 6 MW PV project has been built at Masdar near Sharm-el-Sheikh on the Sinai Peninsula.

- Just recently in November 2022 two large PV projects to be built were announced in **Saudi Arabia**, one of 2 GW and another of 2.6 GW with Chinese participation.

In the longer run even larger projects may be built in Mauretania and Morocco to generate green hydrogen from solar PV.

Sharm-el Sheikh: The author with Dominique Campana from Paris in front of a Christian monastery under Mount Sinai, in the background. The monastery is one of the oldest of Christianisme that survived in the centre of the Islamic world.

Chapter 3

Climate Change and PV

The National Oceanic and Atmospheric Administration (NOAA), the US agency that was long ago the first to alarm the world about climate change, warned in a new report in mid-2022 that the content of greenhouse gases (GHG) in the atmosphere keeps increasing all the time. By October 2022 it stood at 415.95 ppm + 2.3 ppm per annum. (It is measured at the observatory in Hawaii, far from any other pollution; I visited the place once myself.)

The GHG emissions originate from the burning of fossil fuels, mostly for the world's energy markets. The climate change that is a consequence of these GHG emissions is well with us by now.

Earth's surface temperature is 1.2°C higher now than previous averages of the 19th century. The temperatures are increasing with higher latitudes and ice melting affects the Arctic regions. Other more Northern countries, in Europe for instance, enjoy becoming wine growers. There are also regions in the temperate climate with more pleasant living conditions.

But in general, impacts of climate change are devastating: Conflagrations in 2022 in California, Spain, France and many places around the world; Inundations like the one that left half of the whole country of Pakistan under water, floods in Australia, the Sahel countries in Africa, and in regions that had previously been dry, or devastated by fire. On the other hand, there had been severe droughts, storms.

Solar Euphoria: The Rise of Photovoltaics to the Top
Wolfgang Palz
Copyright © 2023 Wolfgang Palz
ISBN 978-981-5129-00-7 (Hardcover), 978-1-003-43866-3 (eBook)
www.jennystanford.com

"The Lancet Countdown on Health and Climate" of **October 2022,** the seventh published since 2015, made it clear that the climate crisis is not only dangerous for the environment in which we live but also for man directly. The Lancet study was led by 99 international medical experts from 51 institutions, among them the World Health Organization (WHO) and World Meteorological Organization (WMO).

The report says it clearly "The climate crisis kills us". The study authors call for a response centred on health.

The report warns that in 103 countries extreme heat is now more frequent. As a consequence people suffer from food insecurity, in 2022 some 100 million people more than 10 years ago. Extreme drought grew by 29% in the past 50–60 years. Unusual heat is a direct thread to health. Malaria and dengue infections are increasing. Better air quality and less exposure to the fossil energy could have avoided 1.3 million deaths just in 2020.

And the report underlines that the countries are themselves part of the problem by subsidising fossil fuel production and consumption by $400 billion net in 2019.

The **WMO in Geneva** found in a separate report that weather and climate events are at the origin of hundreds of deaths. In 2021 half a million people suffered from these events. Economic damage was over $50 billion. Two-thirds of all these catastrophes were unusual floods and storms. Europe is concerned in particular. In no other continent temperatures increased so fast in the last 30 years. Glaciers in the Alps lost 30 meters in thickness in the last 25 years.

Remember the **UN Conference COP 21** with its **Paris Treaty** and its aim to reduce GHG emissions so that Earth's temperature increase is limited by mid-century to +2°C or even better, only +1.5°C. The Treaty had been signed by 193 Parties (countries) in 2016, but by 2022 most of them had not followed suit. On the contrary, investment in fossil fuels steamed ahead!

Considering that we got already 1.2°C global temperature increase by now and the consequences are devastating as mentioned before. What strange regime are we following then? The Paris Treaty is suggesting even higher temperatures than the 1.2°C we got. At the time of COP 21 it was the best compromise possible. But still, it was not a good compromise: PV and the renewables, the ways to solve the climate problem were not even discussed at the Conference; while nuclear stood on the bridge for a renewal.

The Paris Treaty goals mean even more terrible weather events than those we have already. And worse, the increase up to +1.5°C or +2°C are not taken seriously by most countries... It looks as if a real climate disaster is looming and no route to escape!

The Fossil Fuel Finance Report 2022 found that in the 6 years since the Paris Treaty was adopted the world's 60 largest private sector banks financed fossil fuels with $4600 billion, $4.6 trillion. The report's authors, among them Rainforest Action Network, called it also "**The Banking on Climate Chaos Report 2022**". The report underscores the stark disparity between public climate commitments of the world's largest banks versus the reality of their largely business-oriented one, the usual financing in the fossil fuel industry. $742 billion were invested in the fossil fuels in 2021 by private banks, more than in 2016 when the Paris Treaty was signed. **Climate Action** found that only 7% of the global banking energy finance between 2016 and 22 went on renewable energies. In 2021 the 60 banks profiled in the Fossil Fuels Finance report funnelled $185 billion into the 100 companies leading expansion in the fossil fuel sector. Over the last 6 years **fossil- fuel financing was dominated by the 4 US banks JPMorgan Chase, Citi Group, Wells Fargo and Bank of America**. They account for ¼ of all fossil fuel financing. The report underlines how banks joined the "Net-Zero Banking Alliance", part of the Glasgow Financial Alliance for Net Zero in 2021 while simultaneously financing oil and gas expansion companies and helping to lock the planet into decades of climate disasters.

Canada is the second largest country globally after the United States subsidising the fossil fuels. Following the above report the investments of the Royal Bank of Canada in the fossil fuels went from between 2020 and 2022 from $19 billion to $39 billion.

The top 20 companies supported are responsible for more than half of fuel development and exploration. Each of the top

10 bankers supporting them had formally agreed to net zero emissions by 2050—green-washing on a large scale.

Tar sands saw a 51% increase in financing by US and Canadian banks from 2020 to 2021 up to $23.3 billion. **$62 billion went to fracking in 2021**; Arctic oil and gas got $8.2 billion that year. Enormous LNG infrastructure projects are pushed through. $52.9 billion went to offshore oil and gas in 2021. Big banks led by the Chinese Everbright Bank provided $17.4 billion for coal mining.

And the result? **Instead of decreasing as signed by all in 2016 in the Paris Treaty, all GHG emissions rose in 2021 by 6%, the highest increase ever!** WMO announced end of October 2022 that next to CO_2, methane, a strong GHG, made a particular jump in atmospheric concentration that year. In the run-up to the COP 27, a new Conference of the Parties (countries) held in November 2022 at Sharm el-Sheikh on the Sinai Peninsula in Egypt, the head of UN Climate raised the concern that the world is currently engaged on the way to increase global surface temperatures by +2.5°C and not the +1.5% "needed to alleviate climate change". (Overlooking that in reality temperatures should not go up at all, but down.)

UN experts underlined that by 2030 emissions must go down by 45% compared to 2010. But they are skyrocketing by 10.6%, the trend in 2022. In its report of 2021/22, IPCC, the International Panel on Climate Change had warned that time is running out to keep the globe liveable.

Not only the UN and the world's energy specialists are highly concerned. Our societies at large are alarmed about climate change and its consequences on living conditions. In opinion polls today it comes on top of people's concerns.

The polluters and the industry responsible for the ever-increasing GHG emissions are known. On top of the list come the providers of our daily electricity and energy at large. Without them we might be living in the dark and our economy come to a standstill. You will not complain about them for that. What you can complain is that they are not doing enough to turn away from the fossil fuels and on the contrary, as shown in some examples above, to strengthen investments in them. And in the expensive and dangerous nuclear power.

There exist many others in business and finance that are mobilising against climate change. Since 2019 there is "Climate Pledge" committed to net zero emissions by 2040, 10 years earlier than demanded in the Paris Treaty. It was co-founded by Amazon and has some 300 members, such as IBM, Microsoft, Best Buy, PepsiCo, Siemens, Unilever, Verizon, Visa. The companies are dedicated to enlarge the share of the renewables in electric consumption. Employing PPA contracts are specific means to achieve their pledge.

In October 2022 some 318 financial institutions and multinational firms weighing in total $37 trillion ($37,000 billion) addressed leading emission companies asking them to better respect the goals of the Paris Treaty on Climate Change of 2016. Nearly half of the 3000 companies on the All Country World Stock Index MSCI have no credible targets yet, they are told.

They did not yet adhere to the "**CDP Science Based Targets initiative SBTi**" it is said. That campaign is coordinated by CDP, a global NGO that runs the world's environmental disclosure system for companies, cities, states, and regions. Founded in 2000 and working with more than 680 investors with over $130 trillion in assets. It motivates companies to disclosure their environmental impacts, reduce GHG emissions, safeguard water resources and protect forests. Over 14,000 organisations around the world disclosed data in 2021 through CDP, including more than 13,000 companies worth over 64% of global market capitalisation.

Companies with targets approved by the **SBTi** typically cut GHG emissions by 8.8% a year, well above what is needed for the 1.5°C reduction path defined by the Paris Treaty. The CDP campaign of 2022 is 30% bigger than the year before. Among the large financial institutions and multinational firms in total worth $37 trillion in assets and spending power issuing the call this year one could mention, for instance, the **European Investment Bank, UBS, Allianz, Global Investors, Crédit Agricole, or Nomura Asset Management, Yamaha, Astra Zeneca.**

On the other side, the list of 1000 companies targeted are more from Asia and the USA and less European, the latter having been so far better in those target settings.

But very little commitment from the big power companies so far to the **CDP STP initiative!** Only some 50 power companies from 20 countries have joined and they come mostly from the renewable sector; they are not from the fossil fuel power sector.

At last, financial support to PV is recorded in "**Global Climate Finance**". Sorry, we have to bother the reader again with a jungle of organisations active in the field of energy and the environment: The "**Global Climate Finance**" report was published in its 2021 edition by the **"Climate Policy Initiative" (CPI)**. CPI is an analysis and advisory NGO. Its mission is to help governments, business and finance institutions. It is a leader in tracking sustainable investments in transition to a low-carbon, climate resilient economy.

Global climate finance, following the report, almost doubled in the last decade with a cumulative $4.8 trillion or $480 billion annual average.

It is important to note that Photovoltaics and a few other renewable energies continued to be the main recipient of mitigation investment, says the report. They were mainly financed through private capital, reflecting growing commercial viability. The report notes that PV and some offshore wind power has been transformed into an established and competitive sector with a 7× higher return on investment than fossil fuels. Public sector support was crucial for scaling investment by supporting technology cost reduction and providing incentives.

Continued fossil fuel support remains a barrier to achieving global climate goals. The total fossil fuel subsidies in 51 major countries were 40% higher than the total global investment in climate finance from 2011 to 2020. The report of CPI notes that this is alarming as the fossil fuel subsidies are only a part of the emitting activities.

For the rest of the decade, the report warns, we need at least $4.3 trillion in annual finance flow per year to avoid the worst impact of climate change: the $480 billion, the last annual finance increases are not on track to meet a +1.5°C global warming scenario.

Bloomberg NEF issued a report on the global investment in the low-carbon energy transition on January 26, 2023. It found that from the record amount of $1.1 trillion spent in 2022, $495 billion funding from businesses, financial institutions, governments and users went on the renewable energies. $24 billion went on new PV factories, most of it in China.

Investments matched for the first time those on fossil fuels.

Chapter 4

Global Energy Transformation. A World of PV?

Electricity and its emissions in generation and consumption come first in the world's total final energy consumption. In 2021, fossil fuels and their polluting GHG and other emissions were still the leading producers of it with 62% of the total. PV and wind power together came in for some 10%. As newcomers, both gained already a respectable market share as they entered world markets only since the year 2000.

We have seen that experts in energy and climate see an urgent need to increase massively the share of those in global electricity generation. PV is called to come on top of future market expansion among the renewables' implementation. It did so over the last decade when it became a relevant global player passing for the first time in 2019 wind power in global power installation.

The question arises then what could be PV's future role in displacing in particular the dirty fossil fuels out of the electricity market and preserve climate. By now in 2022 the fossils stand for 16,600 TWh of total yearly generation and PV for just 1150 TWh. Global PV grew another 250 TWh in 2022, but the electricity from fossils, too, increased by 1200 TWh between 2019 and 2021. It looks like an impossible challenge to ban part of the fossils by 2050 as they are even increasing their part by now.

The International Energy Agency (IEA) of Paris, in its World Energy Outlook published in October 2022, even though

Solar Euphoria: The Rise of Photovoltaics to the Top
Wolfgang Palz
Copyright © 2023 Wolfgang Palz
ISBN 978-981-5129-00-7 (Hardcover), 978-1-003-43866-3 (eBook)
www.jennystanford.com

predictions by nature are speculative, wants to be optimistic. It notes that policy responses in the USA, the EU, and in Asia are fast tracking the emergence of a clean energy economy. It thinks that clean energy investments grow (the IEA is not more specific and sees even nuclear together with the renewables) to more than $2 trillion by 2030, a huge opportunity for growth and jobs.

(By the way, for those not very clever politicians who did not yet get it that nuclear is not only dangerous and expensive but also unreliable: The price of electricity jumped in 2022 in Central Europe to unseen highs; in 2020 the MWh had cost €40, in 2022 its cost went up to €1000. A reason is that half of the 56 atomic reactors in France were not operating—electricity supply in Europe became tight as France's neighbours must help to fill the gap. Germany had a surplus of production and exported a lot of it. Much of the exported electricity was from PV and other renewables; they supply already half of total electricity generation in Germany, Europe's leading economy.)

There is further encouragement for PV's global markets in the future: Following **Ember** in London, renewables met in 2022 all the growth in global electricity demand, a third of it by PV. Consequently, global CO_2 emissions from the power sector in 2022 were not increasing. Ember analysed that in the EU demand of coal rose sharply because of restrictions of gas from Russia, but at the same time these rises were offset by falls of electricity use from coal in the United States and China. (Actually not exactly all the growth in electricity demand was met by the renewables, as China was hit hard by a heat wave in the summer 2022, decreasing hydro production that it compensated by some more electricity from coal and gas.) A year before in 2021, growth came only by 50% from the renewables in front of a jump in electricity from the fossils that year. Now in 2022, global generation from coal and gas remained almost unchanged.

Just for the first half of 2022, always following Ember, PV and wind power prevented an increase of 4% of electricity from fossil sources. That avoided $44 billion in fuel costs and 230 Mega-tonnes of GHG emissions. So, in China fossil-fuel electricity fell 3%, in India it rose, but only by 9% instead of 12%, and in the United States it slowed from 7% to only 1%

increase. In the EU that became so coal hungry in 2022, PV and wind power prevented the fossils to rise by 16% instead of 6%, the actual rise.

In total, the world's electricity demand grew by 3% during that time. 77% of it was met by PV and wind power. Those met 81% of the growth in the United States and even 92% in China!

We will restrain at this point to make here, too, projections about future global markets of PV, in particular at a time of war in Europe and the political tensions that follow, or with the economic tensions between the United States and China, that is already called by some people a cold war; and worrisome for the PV markets given China's leading role as a supplier. Let's just fix the orders of magnitude:

The power capacity installed worldwide was 7.76 TW (7760 GW) in 2020: 4.4 TW for fossil fuel combustion and 0.7 TW of PV power, not so much less actually. In 2022 PV power came to 1.1 TW and some 4.7 TW for the fossils.

In 2023, PV comes 1.3 TW, = 16.2% of the total power of some 8 TW installed.

So, as stated earlier:

The 4.7 TW of fossil fuel capacity in 2022 generated 16,600 TWh, as the 1.1 TW of solar generated 1150 TWh. It is well understood that electricity generation from PV in kWh per kW installed is more restrained because of the limitations of solar irradiance. PV generated 1150 TWh. The mean running time per year is longer for the fossil-based ones: 3530 hours and for PV "only" 1050 hours. Of course, these hours for PV are equivalent hours of full power. PV operates a lot more hours in the year but at lower power following the meteorological conditions.

For future markets one may first consider the growth rate of the world's PV markets in the past; it was on average 32% from 2010 to 2021. A similarly big one could continue for a while but not forever obviously. Currently, in 2022 we have achieved a power capacity of 1.1 TW in PV with a growth of over 200 GW this year. At some 30% growth rate 3 TW from 1 TW now were achieved in the 7 to 8 years to the end of the decade. That would already get PV with 3 TW in terms of power capacity not so far behind the almost 5 TW of the capacity of fossil fuel power by 2030 or so. But more is needed to increase solar generation. Supposing growth would even continue at high rate, it might not look impossible to achieve PV power capacities high enough to displace a sizable part of the fossil fuel capacity. This will be helped, as experts think, by electricity from carbon slowly disappearing on its own right as it is the world's biggest polluter, not only for GHG. But there is also an important rise in electricity consumption to come as the world is ever more moving to informatics, electric cars, heat pumps, and appliances. That increase will have to be covered, too, by clean and homemade solar PV.

Everything is possible given current market strength of PV. There is obviously the need but also the readiness to invest huge sums in PV as they are highly profitable: in 2022 it was well over $200 billion of private investment. In the future it might eventually become almost $1 trillion a year.

In a recent **"Solar Futures Study"** for the US federal administration, the US National Renewable Energy Laboratory (NREL) found that the benefits of de-carbonisation far outweigh additional costs incurred. In its analysis, for the case of the United States, cumulative power system costs from 2020 to 2050 are $562 billion (25%) higher. However, avoided climate damages and improved air quality more than offset these additional costs, resulting in net savings of $1.7 trillion.

And there is PV's twin, wind power. It is a strong contender in the clean power markets, too, though it recently lost a bit of steam; **opponents of wind power are not disarming, on the contrary.**

Opposition to large-scale deployment of PV, as far as it is concerned, is relatively rare. No wonder, as globally almost half of all PV plants are building integrated, and many of the other half provide attraction and even beauty to otherwise bare or set-aside ground.

But when it comes to PV deployment for providing major parts of the electricity supply of a region or a country, the questions do arise whether the areas to be exposed to the Sun may be prohibitive in size. But they are not. In the NREL report "Solar Futures Study" mentioned above, for the United States, **land availability does not constrain solar deployment:** In 2050, ground-based solar PV require a maximum land area equivalent to 0.5% of the contiguous US surface area. This requirement could be met in numerous ways, it is said, including the use of disturbed or contaminated lands unsuitable for other purposes.

For building-integrated PV, it was calculated for Germany already some time ago that the roofs, façades, and other structures available for PV installation would in total suffice to supply up to five times national electricity consumption.

Another possible problem when considering large-scale deployment on a regional or national scale is the Sun's **intermittency.** In the last few years, **battery storage** on a local scale has become widespread. This development had indeed benefitted from the pioneering use of efficient and lightweight lithium batteries in electric cars. When envisaging the deployment of PV on a multi Terawatt scale, that means a world of lithium together with the world of PV. This looks problematic.

Since a few years we have seen a new interest in "**green hydrogen**". An interesting option for clean fuel generation in general and clean power for storage. We will come back to it later in a separate chapter.

Eventually, there is the option of a large-scale combination PV generation with electricity from wind power. The deficiency of solar energy in the winter months is balanced by more strength of the wind at that time. It was studied in detail for Germany and Europe by the German meteorological agency Deutscher Wetterdienst (DWD). In 2019, Frank Kasper and co-workers published a thorough and detailed study on the subject: **"A climatological assessment of balancing effects and shortfall risks of PV and wind energy in Germany and Europe"** (*Adv. Sci. Res.* **16**, 119–128, 2019).

Wind speed and solar radiation are not suitable for direct comparison because the non-linear relation between the meteorological parameters and potential energy yield. The DWD study uses the **capacity factor (CF)** to make wind speed and solar radiation comparable. The CF is unit-less. It gives the ration between the actual energy output over a period of time and the maximum possible energy output over that period.

The results are encouraging. The following figure shows the cycle of the CFs for wind and PV for Europe.

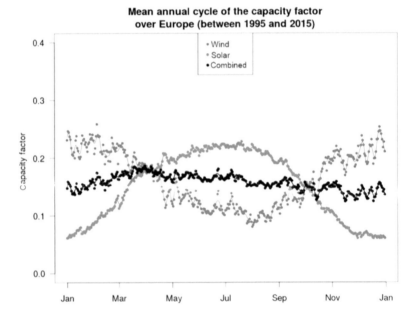

Global and direct solar radiation data have been derived from meteorological satellites over Europe. For calculating the CF of PV, a c-Si module was assumed, inclined South at a 25° inclination and 10 m/s wind speed. CF comes as its variable performance over the year in comparison to standard test conditions of PV at 1 kW radiation input and the module at 25°C.

The wind CF was performed by using the power curve of a large 7.5 MW wind turbine and a hub height of 116 m for it. Wind datasets of several meteorological parameters were gained from radio probes, aircrafts, wind profilers, and surface level data with a spatial resolution of 6 km.

The CF was calculated for each hour of the year.

The seasonal cycle of the capacity factors shown in the figure illustrates the complementarity of PV and wind power electricity over the year. The black line is the average based on the assumption of equal share of capacity for both. The number of events, or days, when the capacity factor remained under 10% is reduced to just a few over the year—to be bridged by some storage.

Similar studies have been performed by other groups demonstrating complementarity of PV and wind power for China, Brazil, West Africa, or Canada.

PART II

REVIEW OF A FASCINATING PV EXPANSION IN THE 20TH CENTURY

Chapter 5

PV's Scientific Beginnings since Becquerel in Paris (1839)

Edmond Becquerel discovered the Photovoltaic effect in 1839 in Paris. He was then just 19 years old and worked in his father's lab at the Museum National d'Histoire Naturelle right in the centre of Paris, where it exists until today. Later he described his discovery like this "...observes an electric current when one exposes unequally to solar radiation two sheets of silver or gold in an acid, neutral, or alkaline solution..." He was aware of the importance of his discovery and later in 1867 wrote the book *The Light Its Causes and Its Effects* (in French).

Later, in the building, his son Henri Becquerel discovered, before the Curies, the radioactivity effect. It was the radioactivity of uranium.

Later the Curies became very popular in Paris with their work that followed Henri's. There is an Institute Curie, a Museum Curie, Curie books, and Curie streets all over France. Nothing named after Becquerel—but still, there is the Becquerel unit for radioactivity.

And only in 2021, Daniel Lincot, Director at CNRS, organised the fixation of a memory plate in honour of Edmond, the discoverer of PV, at the entrance of the laboratory building in the Jardin des Plantes in Paris.

Solar Euphoria: The Rise of Photovoltaics to the Top
Wolfgang Palz
Copyright © 2023 Wolfgang Palz
ISBN 978-981-5129-00-7 (Hardcover), 978-1-003-43866-3 (eBook)
www.jennystanford.com

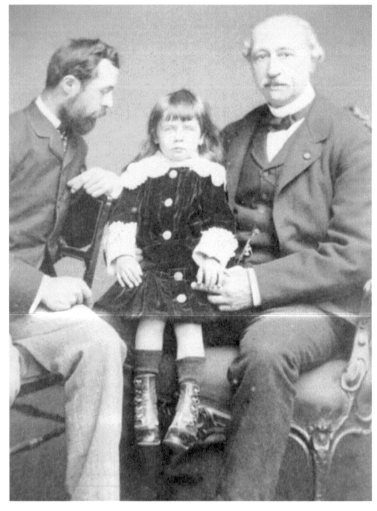

The Becquerels. (Right) Edmond, the discoverer of the PV effect. (Left) His son Henri, the discoverer of radioactivity. Picture credit: Loic Babo, Les Génies de la Science.

However, the English reminded well what had happened in 1839: in 1989, for the 150th anniversary of the discovery of PV, my friend the late Prof. **Bob Hill** of the University in Newcastle in England got the BBC to show a program on Becquerel. Thereupon I decided on behalf of the EU to create a Becquerel Prize for Merits on PV development. It is regularly remitted until this day in a formal sitting at each of the European Conferences of PV, EU PVSEC, the series that I had created previously in 1977. Dozens of pioneers have been honoured since then with this Prize. The first one was late **Baron Roger van Overstraeten** (1937–1999) from Leuven in Belgium, a PV pioneer and a friend who passed away too early. He was in 1984 the founder of IMEC in Leuven, in Flanders, today one of the leading centres of microelectronics in Europe and the world.

Selenium, not silicon, was the first solid material used for a practical solar cell. Its spectral response is less adapted to solar irradiation than that of silicon, but technology wise it is much easier to produce. The technology of semiconductor-grade silicon is complicated and, as we will see later, became ready only in 1954.

But before the first PV effect was discovered in selenium, the English **Wilboughly Smith** accidentally found the **photoelectric effect** in this material. The difference with PV is that photo-electricity does not involve the creation of a voltage with the current under the effect of illumination. It is suitable as an **optical sensor**, but not for the generation of energy.

Selenium optical sensors survived until late in 1960s as light meters on German cameras.

A first **PV effect on selenium** was discovered by **Adams and Day** in London in 1876. It was also found accidentally as they had not built in purposely a diode that is essential for a PV effect.

The real thing happened in the **United States in 1883** when **Charles Edgar Fritts** built a **first selenium solar cell**. It had a semi-transparent gold contact layer that provided the diode effect. It is reported that Fritts achieved an efficiency of just 1% with his cell. Interestingly he is quoted as saying. "**We may see the photoelectric plate competing with fossil fuel plants…**" Fritts sent his cell to **Werner von Siemens** who showed it at the Royal Academy of Prussia in Berlin. Everybody was impressed. Siemens is quoted saying "**first time the direct conversion of energy of light into electrical energy**". The first solar cells had too low an efficiency and were too expensive for practical applications. **But still the first electric car with selenium solar cells was built in the United States in 1912.**

The world's first practical solar cell with 6% efficiency was built at Bell Labs in 1954 by Daryl Chaplin, Calvin Fuller and Gerald Pearson, assisted by **Morton Prince** (the latter passed away 98 years old, which saddened many of us in the summer of 2022). **Bell** was a very successful Laboratory of **AT&T.**

The cell was a mono-crystalline silicon solar cell. In the meantime, Bell Labs had accomplished the task of realising ultra-pure silicon crystals, ready for doping to generate a diode. *The New York Times* wrote, "Solar cells will eventually lead to a source of limitless energy from the Sun". But for the time being first applications were no better than dollar bill changers and toys. Later on, PV became of strategic importance since the early satellite applications in the race between the United States and the Soviet Union with the Vanguard and Sputnik satellites in 1958.

The company "National Fabricated Products" took the license of Bell's Patent on the silicon solar cell. In 1956, this firm was acquired by the small newcomer **Hoffman Electronics**, founded in the 1940s; they owned then the license for the first silicon solar cells. Hoffman produced the first solar cells for the satellites. After a study by Harvard business school from 2012, these first cells had an efficiency of less than 1%. But by 1960, Hoffman had already achieved 14%.

Others got also involved. A certain A. Mann formed **Spectrolab**, together with **Heliotech** in 1956. They developed the new PV for the **Air Force**. They were joined by Eugene Ralph, a pioneer still active in PV until now in 2023, and Martin Wolf—more on Wolf later. Still other entities got involved in the new promising PV, like **RCA in Princeton**. The German-US pioneer Martin Wolf had been working with all these new companies, ending eventually up at the RCA—like Rappaport, the founder of NREL.

Right from the early days when the understanding of semiconductors had sufficiently developed, one knew that many elementary semiconductors and compounds are suitable for PV solar cells. As a possible alternative to the just developed silicon cells came up the **"thin-film" solar cells** and modules. For their inherent properties, silicon cells must be at least 100 micron thick. However, with other materials, achieving a factor 100 less is possible: the thin-film cells. Later more on them.

Another example of a suitable material is **GaAs** that found a special market in satellite applications—because of its high cost, other markets are hardly accessible. A GaAs hetero-junction solar cell was first developed in 1970 by the late **Zhores Alferov** in Leningrad. At the same time he developed from the same material a first **semiconductor laser** (he was an exceptional personality who got both the Lenin Order and the Nobel Prize for his many achievements). *I met with him in 2008 in what now had become Petersburg again, together with my German friend Prof. Klaus Thiessen from Berlin.* Zhores Alferov passed away in March 2019.

By now efficiencies up to 28% have been demonstrated on GaAs solar cells.

In 1976, **David Carlson** and **Christopher Wronski** at RCA in the United States created the **amorphous silicon** (a-Si) solar cell. Previously **Prof. Spear** in Scotland had found that the material is a semiconductor when it is given a 1% hydrogen content and can be N- and P-doped. The little pocket calculators from Japan that had been popular since the 1970s were powered with such little cells. Towards 2010 some 15% of the world's PV market share was a-Si. These cells have disappeared from global markets as one was unable to overcome the inherent stability problem in the material due to the excessive mobility of the small hydrogen atom in it.

The main producer had been **Sanyo** in Japan. Saving what could still be saved, Sanyo that was acquired by Panasonic in 2009, developed the **HIT cell**, described in a previous chapter. In the HIT cell, a hetero-junction is produced by the contact of an ultra-thin layer of a-Si on crystalline silicon: nanotechnology on solar cells.

Next to crystalline silicon, **CdTe** solar modules are nowadays the most successful ones on the markets. The original pioneer was the late **Dieter Bonnet** in Germany. For his achievements he received the European Becquerel award and a street was named after him in his hometown. Dieter established the basic technology of the cell in 1971. It was the same CdTe/CdS hetero-junction that is still commercialised today. Since 1999 **First Solar** is the market leader for these cells and modules.

And there are **CIS** and **CIGS** thin-film solar cells and modules. They had first been proposed in 1971 by **Prof. Loferski** at

Brown University in Providence, USA. After much international research on it, **Solar Frontier** in Japan had a production of it running. It seems that they gave up production in the late 2021.

There is actually a lot more to be said about the fascinating times when PV started to put its neck out. See our book *Solar Power for the World. What You Wanted to Know about Photovoltaics*, published by Jenny Stanford Publishing, Singapore, 2014. In the book, 40 international solar pioneers report on their early work. **Mort Prince**, one of the fathers of the first silicon solar cell in 1954, wrote about the book: "I want to congratulate you on the quality of the book and the quantity of information that you were able to incorporate into it. Just reading the first 50 pages or so I found so much information that I was not aware of... And thanks for the tremendous effort in producing such a fine volume".

Chapter 6

1954: Start-Up of PV's Industrial Development

The semiconductor silicon world started in 1954, for PV and also the micro-electronic chips, and the informatics world that is ours today that followed. The first industrial solar cell and the first industrial transistor, both on crystalline silicon, were created and demonstrated that year.

In view of the enormous market perspectives for integrated chips and solar cells, the first task one had to achieve was the production of ultra-pure silicon material. The market for it grew indeed from the very beginnings in 1954 to over a million tonne by 2022–2023, most of it for solar cells. Details were given in a previous chapter in PV technology. Originally **Wacker** in Germany had developed the Siemens process for purification, and **Werner Freiesleben** (1929–2013) had for this purpose set up the branch Siltronic. With the recent explosive growth of the global PV markets, new companies in the United States and in particular in China have gained the leadership.

Freiesleben was convinced of the coming of a Solar Age with PV. Later he transformed the Wacker branch Siltronic into **Heliotronic.**

As mentioned in the chapters on PV technology already, single crystals—mono-crystalline silicon that dominates today in the 2022s the markets again—was first developed in 1916

Solar Euphoria: The Rise of Photovoltaics to the Top
Wolfgang Palz
Copyright © 2023 Wolfgang Palz
ISBN 978-981-5129-00-7 (Hardcover), 978-1-003-43866-3 (eBook)
www.jennystanford.com

by the Pole **Jan Czochralski.** Casting of poly-crystalline silicon was developed in Germany by **B. Authier** from Wacker and **H. Fischer** from AEG-Telefunken. In 1978 both received the **Walter Schottky Prize** for this work.

The cutting of the hard silicon blocks, be they poly-or mono-crystalline, is not a trivial task. To obtain wafers of minimum thickness of less than 200 micrometres without losing much sawing powder, the kerf loss, **Charles Hauser** in Switzerland with the support of my good friend **Guy Smekens** in Brussels developed the **diamond wire saws.**

To develop the PV technology that was in the early days tailor made for satellite applications, the first PV market, **Joseph Lindmayer (1929–1995)** and **Peter Varadi (1926–2022)** (both early immigrants from Hungary after the first revolution there in 1956) made the essential switch from PV for space to **PV for terrestrial applications**; those have to be cheaper by at least one order of magnitude. For satellites, cost is no criteria, it is rather low weight and durability. Lindmayer and Varadi quit in 1973 their employer **Comsat Lab**, in those days the operator of the international satellite corporation **Intelsat,** to form from scratch **Solarex,** a world pioneer for the deployment of PV for power generation. (Intelsat, for the world's first communication satellites, first developed for the pleasure of Olympic games transmission: The first ones, only some 60 years ago.)

The story of the **first PV developments in the United States** was reported by someone who played a key role in it, another friend of mine, the late **Martin Wolf (1923–2004),** from **RCA's Astro-electronics division at Princeton:** "Solar Cell Development", p. 113 in *Solar Electricity*, proceedings of a PV Conference held in Toulouse, France, in March 1976, of which I served as the secretary general. Martin Wolf had emigrated from Germany in the early 1950s. He was a US solar pioneer who had worked with the early PV industry in California, Hoffman Electronics and Heliotech for **NASA.** When I first met him in Princeton in a private meeting in 1969 or so, he showed me on a few schemes the whole future of PV as the American pioneers had imagined it by that time: it had everything, from central

PV power stations, PV building integrated, residential, commercial, industrial. The large-scale implementation as it is "en route" today. In that meeting Martin Wolf changed myself from Saul to Paul on PV.

Just to finish the story of Solarex attempting to make PV efficient and cheap, Lindmayer first developed the "**Violet cell**". It had better efficiency achieved by three changes, a better anti-reflection coating, a thinner diffused region on the surface and a finer grid structure. One step further he achieved—and that was already in the early 1970s—an 18% efficiency with the "**Black cell**". Here the surface was additionally shaped into a cone-like structure.

Later came into the picture **Dick Swanson.** He was in 1985 the founder of the company **SunPower**, another pioneer in large-scale development of PV in the United States. **Pierre Verlinden,** a Belgian PV scientist who worked later also with the company Trina in China, developed the totally black back-contact cell when working with Dick Swanson. The cell has both contacts, positive and negative on the back, the Interdigital Back-Contact (**IBC**) cell.

Chapter 7

US Pioneers Had a Dream

Imagine the power situation in the United States in the early 1970s. Billions of dollars had already been spent on nuclear power and hundreds of millions on coal power plants. It was planned to develop a liquid-metal nuclear breeder reactor and coal gasification for combined-cycle power plants by 1980. For the former, one projected building one 5 GW breeder reactor every day. In the longer run, nuclear fusion was on the cards of the conventional power industry and their associated National Laboratories. The energy Goliaths had been flexing their energy muscles.

And solar power? It did not count at all. Twenty years after the country had invented the silicon solar cell, its yearly market stayed at less than 0.1 MW in 1973, most of it for space applications. Yet in 2000, again 23 years later, the market reached already 1000 MW of power installed. Fortunately, something happened in between. Solar modules cost $20/W for terrestrial applications in the 1970s and $200/W for satellite power generators. For our home markets the price of a silicon module came all the way down by a factor of 100 to 20 cents/W or so today.

The wake-up call came from Bill Cherry of NASA, it sounded like this: "The large-scale utilisation of solar energy will be a legacy for generations to come, something for all citizens to be proud of and a major step towards cleaning

Solar Euphoria: The Rise of Photovoltaics to the Top
Wolfgang Palz
Copyright © 2023 Wolfgang Palz
ISBN 978-981-5129-00-7 (Hardcover), 978-1-003-43866-3 (eBook)
www.jennystanford.com

our planet both from a particulate and thermal standpoint. For these reasons, the large-scale utilisation of solar energy should be initiated. It just might be the difference between survival or the self destruction of man".

Bill Cherry made this declaration in October **1971** at an assessment conference on the large-scale use of PV organised by **Karl Wolfgang Böer** of the University of Delaware; another immigrant from Germany and good friend of mine who passed away in 2018.

At the same meeting, Prof. Josef Loferski, another one of the early American PV pioneers whom we already mentioned earlier, made the point that automobiles are a reference for mass production. In those days they cost $0.50/kg of car, solar cells came $6000/kg. In terms of mass production similar considerations should apply. And that is what happened since then.

A major event was the following workshop in October 1973 at **Cherry Hill, Philadelphia,** organised by the Jet Propulsion Lab (**JPL**) and sponsored by the **National Science Foundation**. The meeting was considered by many as the cornerstone and start of PV development in the United States, perhaps in the world.

John Francis Jordan, Vice President of Baldwin, the piano company, and another friend of mine, with whom I build in the mid-1970s in El Paso the company **Photon Power** for CdS solar cells, was a great defender of large deployment of PV all over the country, enormous, the American way. He came to Cherry Hill and made the point that the United States had at that time a global capacity of 400 GW of conventional power in place. And new conventional power came at $250/kW mostly financed by private investments. Such cost came as a kind of benchmark against which PV had to compete in future. **Needless to say, by now it did.**

The result of a **panel on silicon cells** was presented by **Paul Rappaport** at that time at RCA in Princeton, a colleague of Martin Wolf mentioned earlier, who became later the founder of **NREL** His panel proposed that by up-scaling, a cost of $0.10/W could eventually be reached at a production volume of 50 GW—and the probability of success were high: there is abundant material cheaply available, theory and technology are well understood, and reliability is proven.

It was an excellent projection. By now some 50 years later, taking account of over 500% of inflation since the time the projection was made, we should stand broadly speaking following the projection at less than $0.50/W; today we are even lower! And as we are today at 200 GW/year installed, we are exactly in the range given by the panel in 1973. Good work.

Only implementation did not proceed the way the early American pioneers had imagined it. They had demanded that the government started financing the necessary investments. "Indecision, fluctuating prices, political rhetoric will not generate the confidence for a privately funded PV program" they said. But what always happens is that governments finance nuclear programs, but not the solar ones. As a consequence, the timeframe wanted by the US pioneers was not realised in the way they had wanted because the government did not follow suit.

So far about what we all owe to the American pioneers for having laid the ground for solar deployment on the large scale we know today.

Looking back, one may remark that the early initiatives in America for PV deployment go back to scientists employed in the space business, like NASA, JPL and Comsat. And they were grouping together in the early IEEE PV Conferences, the IEEE PVSC they created. Exceptions had been Prof. Böer and John F. Jordan. Prof. Böer was a semiconductor scientist; Jordan came from the musical business.

The situation in the United States in those days is better described in my book *Solar Power for the World, What You Wanted to Know about Photovoltaics* (p. 66, 2014, Jenny Stanford Publishing, Singapore):

> After the oil price shock of 1973, renewable energies and in particular PV began to garner overwhelming popular support. The American people became interested in solar energy in a profound way, the significance of which is hard to appreciate today—a "solar tsunami" with shock waves around the globe. It would not be unfair to say that was has followed since pales by comparison to the ardent enthusiasm of those days.
>
> The effort cumulated during **President Carters Presidency (1977–1981).** His Secretary of Energy was James Schlesinger. In

California, Governor Edmond Brown Jr. led the charge for solar. The director of the federal PV program was **Paul Maycock**, an outstanding pioneer.

It's no exaggeration to say that in the latter half of the 1970s many people saw themselves already living in the "Solar age". They were committed to a better quality of life. In May they celebrated the Sun Day. In 1978, US Congress passed an RD&D Act on PV. It called for a doubling of PV generation capacity every year until 1988, with the objective to reach a 4 GW level. The budget commitment was $1.5 billion, an enormous amount at the time (even though it was tiny when compared with what had previously gone into nuclear research). I was invited to give a talk at the opening of one of the IEEE PV Conferences, on the podium with a Congressman.

The **Public Utility Regulatory Policy Act (PURPA)** of 1978 provided regulation that utility rate structures could not discriminate against small producers of electricity and that utilities must buy electricity at fair prices. That same year the White House Council on Environmental Quality projected an installed PV capacity of at least 500 GW by 2020. From 1977 to 1980 more than 1300 small PV systems were built on residences, water pumps for irrigation, etc. They were actually built under the Framework for Federal PV Utilisation (**FFPU**) Program which encouraged federal agencies to get involved. Since 1980, a **tax-credit** of 40% of the purchase price has been offered for PV. Low-interest loans were proposed and the establishment of a Solar Bank was considered.

The federal budget for R&D on PV reached $147 million in 1980. Hundreds of solar-powered buildings were developed, both stand-alone and grid connected, including Lord House in Maine, the Olympic Natatorium in Atlanta, Georgia, PV Pioneers in Sacramento, California, Recycling centres in New York, Georgetown University in Washington DC, APS factory in Fairfield, California, a solar townhouse in Bowie, MD, a Liss House in Fairbanks, Alaska.

The first community relying entirely on solar power was the **Schuchuli Indian Village in Arizona in 1978**. The world's first PV powered neighbourhood came about in 1985 when New England Electric installed 100 kW of distributed rooftop PV systems in the central Massachusetts **town in Gardner**. The first larger stand-alone PV systems of up to a 100 kW were installed among other places, in the National Bridge National Monument Park in Utah, Mount Laguna in California, and in Mead, Nebraska.

Among the American States, **California** had already emerged as a leader. It offered even better tax credits than Washington. In some regions or counties, people voted to shut down nuclear plants, almost every dollar coming out of nuclear went into PV. People were gratified to see that contrary to the nuclear industry's prior warnings, electricity rates remained stable after nuclear plants were closed.

US industry followed suit. New production capacity was built and soon the country had overall 25 MW manufacturing capacity on stream, a five-fold increase achieved in just a few years. Even the energy "establishment" like General Electric (GE) and most of the power utilities that were used to do all their business with the "hard energies" started to wonder whether they should not get involved in PV as well. GE would build a five-acre PV installation for Sea World in Florida.

In 1977, a massive "**block buy program**" was launched by the federal government. Almost 2000 kW were bought and installed by 1980. 10,000 kW more had followed by 1987. A 500 kW concentrating system was installed in Saudi Arabia [its director was **Mathew Imamura** whom we hired later in Europe—see the chapter on the PV pilot plants].

All of them followed the Sun by tracking. The largest plant built in 1983 at Carrisa Plains, California, had a nameplate capacity of 6500 kW and two-axis tracking.

The measured capacity factors of the plants in California, Texas, and Arizona turned out to be as high as 35%. The reason was the good correlation of the Sun's availability with the air-conditioning load.

That's what had happened after President Carter moved into the White House in 1977. He created the Department of Energy, issued a National Policy Act with the PURPA and installed PV panel on the White House, which was later removed by President Reagan. As we know, after Carter's tenure as President came to a close, the United States stepped back from their solar commitment. For Reagan and all Presidents that followed, solar has been a case of business as usual.

John Geesman, formerly with the California Energy Commission, stated on September 2, 2009: "When the Renewable Portfolio Standard was signed into law in 2002, California derived 11% of its electricity from renewable sources. In 2008 that number was 10.6%. Every school kid in California knows that most of that comes from policies enacted when Jerry Brown was Governor 30 years ago."

What moved then first was the German market. A politician, a socialist, Hermann Scheer, a German MP in Berlin has succeeded by action of the Parliament on the government to open the doors to private investment with the FIT. The government in capitalistic America was supposed to finance the investments from public budgets. That's a bit contrary to what you would have expected: that socialists invest from state budgets and capitalists encourage private business. But the United States was catching up later and by now, in the 2020s it is again well ahead of Germany for PV deployment.

Chapter 8

PV Front Runners in Europe

PV in Europe started with the cartoonist **Jean-Marc Reiser** (1941–1983). He started publishing drawings on PV early in the 1970s, they were funny and very informative, he had got always the facts and figures on PV right. With my friends in Paris, we discovered "PV for everybody" through him, (the solar circus coming to town...) although we and not him were the PV scientists. He could also be provocative with his erotic popular drawings, but he was a friend that I admired. Until this day I have the picture of the little solar paradise, a gift of him, in my bedroom. And he was always helpful with drawings for our Conferences etc. Without asking for a penny. Extraordinary, this man.

The year 1973 was another important year for PV. As we have seen,

- there was at Cherry Hill the workshop of the US pioneers of PV;
- there came the oil price crisis that provided an impetus to the alternative energies;
- and at first in July that year took place in Paris at the UNESCO House the International Congress "The Sun in the Service of Mankind".

The **Congress at UNESCO** had been prepared since a long time and came independently from the oil price crisis that had its origin in another war in the Middle East. The Congress has

Solar Euphoria: The Rise of Photovoltaics to the Top
Wolfgang Palz
Copyright © 2023 Wolfgang Palz
ISBN 978-981-5129-00-7 (Hardcover), 978-1-003-43866-3 (eBook)
www.jennystanford.com

been initiated by the French solar community. There is a long tradition in France on the promotion of solar energy. The important French philosopher **Pierre Teilhard de Chardin** (1881–1955), a Jesuit actually, a palaeontologist who lived long time in China wrote in his book *The Future of Mankind*: "since the Palaeolithic and Neolithic age, mankind could always expand: growth and proliferation was the same for him. And now, all of a sudden, emerges in front of us at great speed the wall of saturation. What to do to avoid that human concentration—although social unification is a favourable trend—passes an optimum beyond which all increase of numbers means only famine and suffocation".

A practical example for France's deep solar roots is **Mouchot's** solar printing machine from the 1860s. Other examples are the **Solar Furnaces** in the Pyrenees at Mont Louis and **Odeillo** to generate high temperatures from the Sun's radiation. The leading solar scientist in France was **Prof. Félix Trombe** (1906–1985). He built the furnaces for a comprehensive research in high-temperature metallurgy and chemistry. The research goes on until this day although the furnaces being spectacular constructions in a nice sunny environment serve more as tourist attraction now.

The French solar community of those days was very important and influential, but its interest was focused on the thermal applications of the solar energy: high-temperature research under clean conditions, solar heating of houses—Trombe developed also some passive houses—and small CSP. There was the influential **Jean-Pierre Girardier**, and his company **SOFRETES**, who developed mechanical power machines with thermal solar collectors in the kW range for water pumps in the Sahel region.

But the original organisers of the Congress, scientists of a certain age, had left aside PV; it was "unknown" to them. So had to come in the PV scientist from **CNES**, the country's space agency, in the 1960s newly created by President De Gaulle. The company **SAT in Paris** produced some silicon cells under licence from Spectrolab in the United States. PV R&D was concentrated on thin-film solar cells. PV was managed from CNES headquarters in Paris, and **I was its manager since 1967.**

Consequently, it was decided that I joined the organisation of the above Congress and take over the new part on PV that was added to the program; it even became the most important of it.

The Congress was attended by a thousand scientists from all over the world, the United States, Germany, the Soviet Union, the Arab countries, and many others. **Pierre Auger,** the well-known semiconductor physicist (the Auger effect) opened the Congress. He started by **playing a hymn to the Sun**.

Bill Cherry led the US delegation. Academician N. Lidorenko came from Moscow. I invited **Wernher von Braun** the father of the Space Age. But it was already towards the end of his career, he was ill and wrote a nice letter to congratulate us to work on the **"coming Solar Age"**.

Later I prepared the Proceedings of the Congress for distribution to the Congress participants. On contract of UNESCO, I prepared the report **"Solar Electricity, the coming Energy Source"**. It was later further developed and then published as a book by UNESCO.

The year 1974, the following year, was important again.

First I informed my hierarchy at CNES. They became also convinced of the interest of PV for terrestrial applications. The President and DG agreed to meet their counterparts in important institutions in Paris, CEA, EDF. Even a camping car was set up under the Eiffel Tower to show PV applications

In a press conference in Paris, CNES reported on all the opportunities of PV for residential applications, industrial, commercial, power plants and all the large-scale applications of PV

Eventually, the French Government that year organised a strategic meeting on larger energy independence as a consequence of the oil price shock.

CNES and I were invited to make the point of what PV had to offer for France. The other party next to us to present their offer was CEA and all the nuclear community of France.

Needless to say, nuclear won. Immediately thereafter France went ahead building in a record time a comprehensive park of 59 GW of nuclear power. At first, it was decided to use the technology the CEA had developed in-house. Later on, the French took a license from Westinghouse in the United States.

Decision makers in the French Government thought it was a pity to give simply up on PV. France being influential in Brussels, the capital of the EU, rapidly it was decided to set up an R&D programme on the renewable energies for the European Union. Still the same year the EU Commission started the preparation of such a programme.

The Commission invited all the nine member countries, including the UK that had just joined the EU as well, to send a delegation of renewable energy experts to a first brain-storming meeting. It was organised by the DG **Dr Guenther Schuster** at a nice place in the centre of Milan in Italy. I attended as part of the French delegation. When experts were asked to split into groups for discussion, everybody went to the thermal solar energy groups, and I was left with **K. Krebs** from the JRC at Ispra to form a PV group. Mr Schuster found it strange that not more people joined for PV and so he joined ours. He was basically a "nuclear" scientist, but from thereon he understood PV and we became friends even until he diseased 2011 in Bonn. He became my first DG at the Research Department when **I joined the Commission as an Official in 1977**.

The Renewable Energy Programme of the Commission was decided in 1975. I was since 1977 the head of the solar programme, including next to PV, the priority programme, also solar heating, the biomass, and wind power. Nowadays the EU R&D Programmes managed by the Commission in Brussels are manifold and comprise billions of euros. In 1975, when European Research made its first steps, solar came only second just after an environmental research programme had been decided.

So I had left France, which became an important generator of nuclear energy. This trend was hailed by all parts of society—no matter that the German neighbours did after Fukushima decide to leave nuclear completely. Even though there was an eminent nuclear physicist **Frédéric Joliot-Curie,** Professor at the famous Collège de France in the centre of Paris, who declared in May 1956 already, "I think we must very seriously and immediately get involved in the utilisation of solar energy. It would not be reasonable to see in nuclear energy the only source for meeting the considerable future energy needs of our country".[a]

A clear commitment of France's scientific community in favour of solar energy came in 1978. It was the study **"ALTER, study of an energy future for France based on renewable energies"** issued by the "group de Bellevue" with experts from CNRS, EDF, Collège de France, and the Institute for Agricultural Research (INRA). It was led by **Philippe Chartier,** a Director at ADEME, a friend and when needed expert at our European Programmes in Brussels. The study said that in the perspective of 2050, it would be possible to change progressively in France from an energy system dominated by the fossil energies to a stable autonomous energy system based exclusively on renewable resources, i.e. solar energy in all its forms". More on it later.

Another important voice in France against nuclear—first the bombs, then the power plants—was **"Les Amis de la Terre"**.

The Club was founded in **Paris in 1970** on the example of the **Sierra Club in the United States. Edwin Matthews**, an American living in Paris was the founder together with **Alain Hervé.**

They were joined by important stakeholders on environmental matters in France like **Jean Dorst,** a former Director of the Museum National d'Histoire Naturelle where the Becquerels once had their laboratories, and others. **Brice Lalonde, Dominique Voynet, Pierre Radanne** were important members. **Greenpeace** and the political party of the **"Greens"** had their origin there.

The Club addressed all environmental problems of our time: pollution of the grounds and the atmosphere, the greenhouse effect, waste and residues, lead in gasoline, acid rain, GMOs, the water availability and its pollution, the forests, agriculture and husbandry, transport, food and health, militarism, Third World, NGOs, sustainable development, an urban plan for Paris...

They were the main opponents against nuclear. The French atomic bomb tests in the Pacific (Mururoa 1973); the development of the nuclear power plants in France. They organised

---

[a]*See also*: PV's Industrial Successes since 1954: a short summary of a speech given by the author in April 2022 at the prestigious Collège de France in Paris. Available at: https://www.youtube.com/watch?v=Gn8Q_Ylyl7M (accessed on 6 March 2023).

many meetings of protest with thousands of people in France. The largest was held in June 1987 in Paris, a year after the Tchernobyl explosion: **"For a future without nuclear power"**.

In December of 1978 occurred a black out in Paris and Northern France, no light, no heating, no metro, no train, no lift: The Figaro newspaper reminded, now 44 years later, the reaction of Les Amis de la Terre at that event: "In front of that energy policy of gigantism and lack of foresight we propose a realistic program of saving energy and renewable energies." Nowadays the Club has become international, with 72 member countries, many of them in Africa, and one and a half million members: the **Friends of the Earth International (FoEI)**.

Eventually, a turning point came on 18–19 December 2022 with the adoption of a historical agreement of all the world's countries on biodiversity, to protect our land, the oceans and all living species from pollution. "This Pact of Peace with Nature" has been reached under COP 15 organised by the United Nations and chaired by China, represented by the Chinese Minister for the Environment Mr Huang Runqiu. The agreement's road map aims to protect 30% of our planet via some twenty measures by 2030—today only 17% of the land and 8% of the oceans are protected. It is very ambitious indeed! The Pact gives guaranties to the indigenous peoples, the guards of 80% of the remaining biodiversity, proposes to restore 30% of degraded land and reduce by half the risks from pesticides. To reduce waste of food and combat intrusion of foreign species were among other targets agreed.

There is no time to lose: 75% of the world's ecosystem is already altered by human action, over a million species are endangered of extinction. Half of the global GDP depends on nature and its services.

As usual at these Conferences, the less-developed countries did ask for financial aid to share their part of the coming effort. It was the merit of the Chinese chair to solve also that problem in long discussions.

Today one could write a thick book on nuclear energy. Not funny, it must be a book of disasters, political and technical. Les Amis de la Terre organised the opposition to nuclear when its development had started. Now, that France is sitting on a park of old plants in limbo, discussing how to get over the January peak in demand, and a new nuclear reactor that is a technical catastrophe before it produced the first kWh, there is nothing to oppose, a lot to regret.

With the beginning of the 1980s Europe had got everywhere a wake-up call on PV, in particular in Germany. There, a first PV industry emerged at AEG-Telefunken. Siemens in Munich bought ARCO Solar in California, a disaster that ended with the loss of billions of dollars. Prof. **Kleinkauf** created the inverter company **SMA**, and other Professors (Prof. **Götzberger,** the late Prof. **Bloss**) created respectively the Fraunhofer Institute and ZSW.

Ludwig Bölkow had the merit supporting new initiatives on PV from his personal pocket after he had retired from the aerospace company he had created in Munich after WWII. In particular he supported the ideas of **Reinhard Dahlberg** who proposed in 1986 to deploy PV on large areas in North Africa, 45 km^2 to generate all of Germany's primary energy. His idea was to convert the PV electricity into hydrogen for transport overseas.

In 1987–1988, Bölkow prepared a study to establish the cost of electricity production based on PV on a large scale, employing the technology available at the time. Then a war broke out because of jalousies between the managers in industry and the funding agencies. The study was never published.

And to conclude: **Hermann Scheer,** newly elected member of the Bundestag started his political initiatives for the promotion of the renewables. In 1988 he created **Eurosolar.**

He changed at last PV into a global energy leader. More in a later chapter.

8.1   European Projections for Large-Scale PV Deployment

There have actuality been the following four major ones:

1. The aforementioned study **ALTER** on PV in France by 2050. The "Group de Bellevue" realised it and published it in 1978. The group and the publisher of it, Syros, were registered in Paris. ALTER included a **scenario "All Solar"**. It projected

that in the long run, France could generate all its energy needs domestically and sustainably, relying on solar energy in the first place, like follows:

- A maximum of electricity, in particular for residential use:
  - 53% generated from PV
  - 40% from Hydro, and
  - 7% from Wind power
- 80% of the fuel needs from **Solar Hydrogen**

2. The **Bölkow study** from 1986 on Germany that was torpedoed before publication.

3. The most political and the most expensive was the one of the "**Enquête Commission**" of the German Parliament, the Bundestag in Berlin in 1999/2002.

An overview is given by **Harry Lehmann** in my book *Solar Power for the World...* (2014, Chapter 37, p. 557, Jenny Stanford Publishing, Singapore). MPs and experts from all political parties in Germany participated. I was also a member. The results have been published in a thick report in 2002 (*Sustainable Energy Supply Against the Background of Globalisation and Liberalisation*, Deutscher Bundestag ISBN 3-930341-62-x).

Lehmann, a main author of the report concluded "A sustainable energy supply based on energy saving, efficiency and renewable energies can be realised in Germany by 2050." He continued: "This was the first time that it was stated in a report to a government, to a Parliament that renewable energies are not a niche technology but capable to guaranteeing full supply (100%). At the time this report was published, the share of renewable energies was 6%."

4. A study as part of the EU Commission's marketing programme **APAS** from 1994, was implemented under my responsibility: **Long-term large-scale market deployment (LSMD) of PV in Europe, a study led by Bernard Chabot** of ADEME in France in association with the British ETSU,

the Dutch ECOFYS, the University of Karlsruhe, Germany, and ENEL in Italy. For the year 2030, the study projected for Europe a total installed capacity of 155 GW. Now in 2022 we have already passed 200 GW, a nice work, considering that at the time of the study the EU had not even installed its first GW.

For multi-MW up-scaling of silicon and thin-film solar cell and module manufacturing an APAS study by **BP Solar** in England in association with a multitude of European stakeholders came to the conclusion that for a silicon production line of 500 MW/year, a module cost of €1 was achievable without any technology breakthrough.

Now, some 30 years later, and multi-GW production lines being manifold in China we came not far from this projection either.

8.2 PV Programmes of the EU Paving the Way since the 1970s

As mentioned before, the first EU programme on PV and the renewable energies was decided by the EU Research Ministers in 1975, almost 50 years ago; I took over its head in 1977. From the start, photovoltaics was a major component of it. The aim was to lay the ground for large-scale deployment of solar energy in Europe's future energy systems. That was a time when the global PV market stood at 1 MW. Our actions included first

- A strategic assessment of the technologies and their market potential.
- A resource assessment of the solar potential in Europe. In co-operation with Europe's major meteorological offices in a 10-year effort the Commission prepared the first Solar Radiation Atlas. Later we did the same for Africa. Still later came a wind atlas...
- A systematic effort to quantify the social and environmental costs of energy production and use.
- A clearing house activity on misinformation and misunderstandings, such as the energy payback time for PV modules.

- An observatory of industrial development and research. **In the early 1990s, Europe had the highest global density of cell and module manufacturing.** Fourteen small or large companies were involved in crystalline silicon cells and the whole range of thin-film cells and modules, from a-Si and the newly invented microcrystalline silicon to CdTe, CdSe, and CIS and consorts.

In the EU programme our development priorities were (i) solar cells and (ii) system technology, full-fledged generators that produce electricity when needed. A major effort went on crystalline silicon technology. It was a continuous flow of improvements, no leapfrogging. Highlights in the early 1990s were our projects MONOCHESS and MULTICHESS for silicon. For CdTe we had EUROCAD and for CIS it was EUROCIS.

The global PV market was a tiny 3.3 MW/year in 1980 for commercial power. By 1988 it increased 10 times to 30 MW/year. The market was in the field of consumer products and communication. The market in rural areas was difficult as it lacked financing.

By 1987, the production cost of crystalline silicon PV modules had come down to $4/W with some 40% for the wafer cost. The target by Wacker in Germany then was to reach $0.50/W for the wafer. The cost goal for PV modules was **$1/W** at that time.

In the early 1990s, it was felt that the large-scale implementation of solar energy was approaching in Europe. So we embarked on a new initiative to prepare instruments for integrating the new energies into the various levels of society, to prepare for **a new energy paradigm**. The corresponding programme was called **APAS 1994**, a French acronym for "preparatory and support actions". Funding of €25 million for contractual contributions from the Commission has been specifically made available by the **European Parliament.**

Hundreds of stakeholders from all over Europe participated in a collaborative effort. This massive programme was divided into five sectors. In the following are mentioned some of the highlights:

- *Urban planning:* Towards zero-emission urban development

- *Decentralised PV electricity generation:* Study of large-scale manufacturing of PV. Co-operation with developing countries
- *Eurenet:* European Regions Network for Renewable Energies
- *Financial resources:* Assessment of international finance for the development and use of the renewables
- *Prodesal:* Towards the large-scale development of decentralised water desalination.

In 1993, I set up the **European Solar Council** within the framework of the APAS programme. It was an informal platform on renewable energy, also called "The Club de Paris on Renewable Energy" as its first meeting in 1993 was held, and organised by our friends **Liébard and Civel** in the Louvre Museum there.

It was further set up in accordance with an **EU Council Decision** putting in place an **R&D programme 1994–1998**, which demanded specifically that "actions will be arranged by means of networks, several of which will be linked in a major network for the development of the renewables. It will include among others thematic sub-networks, major European utilities, leading architects and building engineers, specialist research centres, pilot towns, regions and islands". The Solar Council actually linked those EU networks, including:

- EUREC, the EU Research Centres Agency on RE (it has survived until this day)
- READ, the network of architects and building engineers
- CERE, the association of municipalities and towns
- EURE, the network of electric utilities

All these groupings met after 1994 and played a role to make of PV and the renewables the energy giants they are today.

8.2.1 Early PV Demonstration Plants in the United States and the EU PV Pilot Programme

PV Pilot Projects in the power range around 100 kW was a completely new subject in **Europe.** When the EU Commission first announced its intention to support projects in this field, it encountered a tremendous interest from industry and also from the national authorities of the EU member countries.

The purpose of the programme was seen as the necessary complement to PV component development. The PV pilot programme had been outlined already in 1977. It was decided in August 1979 by the Council of Ministers as part of the EU's second 4-year programme that included PV and the renewables.

Details were first published in the proceedings of a **common EU workshop with the DOE of the US Administration held in October 1980 in Sophia Antipolis**, a scientific centre near the Côte d'Azure in France ("Commission of the EC: Medium-Size PV Power Plants", *Proceedings of an EU/DOE Workshop*, hosted by the Commissariat à l'Energie Solaire in Sophia Antipolis, edited by H. Durand, P. Maycock, and W. Palz, published by D. Reidel Dordrecht/Boston).

Pilot plants in the United States were summarised in an overview by **Paul Maycock, the DOE Director for PV** at the time. They can be summarised like this:

- PV projects with utility participation, **flat plate,** built 1980/81:

  - Newman Power Station, El Paso, TX, 20 kW
  - Oklahoma Center for Science and Arts, OK, 150 kW
  - Shopping Center, Lovington, NM, 100 kW
  - Beverly Hill School, Beverly, MA, 100 kW

- PV projects with utility participation, **Concentrators,** built 1981:

  - Sky Harbour Airport, Phoenix, AZ, 225 kW
  - Dallas, Fort Worth Airport, TX, 27 kW
  - BDM Office Building, Albuquerque, NM, 47 kW
  - Wilcox Hospital, Kauai, HI, 60 kW
  - Sea World, Orlando, FL 110 kW

- PV projects with utility participation, **Schools,** built 1980/84

  - Mississippi County Community College, Blytheville, AK, 250 kW
  - Northwest Mississippi Junior College, Senatobia, MS, 200 kW

- ○ Georgetown University Campus Building, Washington DC, 300 kW
- **Residential projects,** built 1980/81
  - ○ Phoenix House
  - ○ Florida Solar Center House
  - ○ Three houses in Hawaii
  - ○ Four prototypes at the NE Experimental Station
  - ○ Several prototypes at the SW Residential Experimental Station
  - ○ Tens of lived-in experimental houses

In the same meeting were presented by **A. F. Ratajczak, of the NASA Lewis Research Center, PV stand-alone systems projects** managed also for the DOE as follows:

- Two refrigerators of 550 W in total installed in 1976
- Two forest look-out towers of 588 W in total installed in 1976
- One Highway Conditions Sign of 116 W installed in 1977
- Four insect survey traps of 372 W in total installed in 1977
- Six automatic weather stations of 592 W in total installed in 1977
- One water cooler of 446 W installed in 1977
- **Two Village Powers of 5.3 kW in total installed in 1978–1979, one at Schuchuli in Arizona, one in Africa (in Tangaye, Upper Volt for water pumping and grain milling)**
- One high-volume air sampler of 360 W installed in 1979, at Liberty State Park, NJ, directly opposite to the Statue of Liberty
- One seismic sensor of 40 W installed in 1980 in Hawaii

The work was performed for **US/AID**. In general, PV systems have proven to be reliable and to require minimum maintenance. Only refrigerators have been a problem due to motor failures.

The socio-economic effects at Schuchuli have been minimal. There has been little, if any, change of lifestyle, but due to the presence of electric lights different hours of walking and eating in the evening.

In Tangaye in Africa, the pure presence of PV has brought status and recognition to the village. Pumped water has affected building of bricks out of mud and personal hygiene in the immediate area. The mill has had a positive effect on the villagers. It was planned then by US/AID to increase the PV array.

This talk was followed by another by **E. L. Burgess from Sandia National Lab. at Albuquerque** giving more details on **grid-connected PV application experiments.**

In 1977, the DOE initiated this program to design, build, and operate grid-connected PV Application experiments for systems of 20 to 500 kW. 150 proposals were received in response to the program announcement. After the completion of the design phase in 1979, nine projects were selected for fabrication and installation. The list of projects has been given in this chapter earlier in the presentation of Paul Maycock.

Prime contractors were in all cases local utilities.

R.A. Whisnant from Research Triangle Institute gave another talk on design considerations for DC or AC PV systems in commercial buildings. The conclusion was that AC systems are preferable for higher efficiency and lower cost.

E. C. Kern from the MIT gave an overview of his Laboratory experimental PV systems. **Lincoln Laboratory of MIT** has joined the DOE's PV program in 1976. Dr Kern gave some details of DOE's residential system development activities that the MIT was managing. There are 19 grid-connected PV systems that were designed and realised.

Battery systems were chosen when the system was remote from conventional utility generation, when utility purchase of excess power is not advantageous. Examples of battery-equipped systems were the following:

- a 25 kW PV system with 90 kWh battery for agriculture at Mead, Nebraska, grid connected
- a 1.6 kW PV system with 27 kWh battery for a Museum at Chicago, grid connected

- a radio station of 15 kW with a 56 kWh storage battery at Bryan, Ohio, grid connected
- a 100 kW load centre with 750 kWh battery at Blanding, Utah, connected to a Diesel generator

The first grid-connected residential system of the US DOE's PV Program began operation at the University of Texas at Arlington in November 1978. A battery-equipped system is far more difficult to control than one without. The latter relies on a stiff voltage reference signal to commutate the power inversion process. Grid-connected systems have power converters which are sized to the maximum output of the PV array, not to the size of the load, thus system design is greatly simplified as the system is decoupled from the load.

The EU Pilot Projects came with the second part of this meeting organised in Sophia Antipolis jointly with the US DOE.

The EU had started its PV system program already in 1977 in the frame of its first R&D Program 1975–1977. The three small projects to mention are a residential 5 kW PV generator in Sophia Antipolis, a 5 kW generator for the Alp refuge "Les Evettes", and a 5 kW generator designed and installed in Berlin in 1979 for a broadcast transmitter.

Ranging from 30 kW to 300 kW and totalling 1.1 MW, the EU Pilot Plants were the first major prototypes intended largely for research purposes, to encourage the development of new concepts and processes and to stimulate the European industries. To a large extent, the original pilot plants served as the springboard for many national organisations and industries to create markets for PV applications within the EU and elsewhere. Three important books have been published on this particular programme of the EU PV pilot plants:

- *PV Power Generation, Proceedings of the Final Design Review Meeting on EC PV Pilot Projects*, held in Brussels 30 November–2 December 1981, by W. Palz, D. Reidel, published by Dordrecht/Boston, 1982, ISBN 90-277-1386-3
- *PV Power Generation, Proceedings of the EC Contractors Meeting*, held in Hamburg/Pellworm, 12–13 July 1983, by W. Palz D. Reidel, published by Dordrecht/Boston, 1984, ISBN 90-277-1725-7

- *European Handbook on PV System Technology*, by Mathew Imamura, Peter Helm, Wolfgang Palz, published by the EU Commission EUR 12913 EN 1992, ISBN 0-9510271-9-0

The following is the list of the 16 Pilot Projects. The first 11 were completed on schedule in 1983 and the five latter ones a year later.

- **Island of Pellworm, Germany,** 300 kW, power supply for a vacation centre, with battery, grid connected and with two wind turbines; leading contractor AEG, Wedel/Hamburg
- **Kythnos Island, Greece,** 100 kW, power supply for a village, with battery, associated with a diesel generator and a wind turbine; prime contractor Siemens and PPC
- **Chevetogne, Province Namur, Belgium,** 63 kW, swimming pool pumps, lighting, with battery, grid connected; leading contractors IDE/ACEC
- **Aghia Roumeli, Crete Island, Greece,** 50 kW, electrification of an isolated village, with battery, stand-alone, diesel back up; prime contractors Renault, PPC
- **Mount Bouquet, near Montpellier, France,** 50 kW, power supply to an FM transmitter of Telediffusion de France, with battery, grid connected; prime contractors Photowatt, Teled. De France
- **Nice Airport, France,** 50 kW, power management and control, with battery, grid connected; prime contractors Photowatt, Chambre de Commerce
- **Fota Island, Cork, Ireland,** 50 kW, electricity to pumps of a dairy farm, with battery, grid connected; prime contractor Univ. Cork, AEG
- **Terschelling Island, The Netherlands,** 50 kW, power supply to a marine training school, with battery, associated with wind turbine; prime contractor Holec
- **Kaw, French Guyana**, 35 kW, electrification of a village, with battery, associated with a diesel generator; prime contractor Renault
- **Hoboken, Province Antwerp, Belgium,** 30 kW, hydrogen production by electrolysis, with battery, grid connected; prime contractor ENI

- **Rondulinu/Cargese, Corsica, France,** 30 kW, power supply to dwellings, a dairy farm, a workshop and water pumping, with battery, propane gas generator; prime contractor Leroy-Somer, EdF
- **Marchwood, near Southampton, UK,** 30 kW, with battery, grid connected; prime contractor BP Solar
- **Tremiti Island, Italy** 65 kW, seawater desalination, with battery, stand-alone; prime contractor Italenergie
- **Giglio Island, Italy** 45 kW, water disinfection, cold store, with battery, grid connected; prime contractor Pragma
- **Vulcano Island, Italy,** 80 kW, village power, with battery, grid connected; prime contractor ENEL
- **Zambelli, Verona, Italy,** 70 kW, drinking water pumping station, with battery; prime contractor Pragma and DornierSol

The designs of all the plants were handled as one unit. For the purpose contractors came to regular design review meeting to Brussels that I chaired myself. It was a common learning process, as never had such large PV projects existed before. Topical reviews that were addressed in common were among the following:

- Design and cost of array support structures
- Wind loading considerations in PV structural design
- Wiring, cabling, and terminations
- Power conditioning and control systems
- Battery systems
- Lightning protection
- Recommended equipment and procedures for the determination of the rated power of PV arrays
- Recommended data monitoring and processing
- Operation and maintenance

The EU Pilot Programme was initiated by the Commission in Brussels in 1979. Its total cost was approximately €30 million, one third of which was provided directly from the Commission's R&D budget through contracts with the international consortia set up for the purpose for the development and construction of these plants. The remainder of the necessary funds were provided from national solar programmes in the EU member countries, the European industry involved and the electric utilities that were contractors.

A particular feature was that it was possible to pool together 16 different plants which are spread all over Europe involving many different technological approaches in one single programme. The programme has become a good example for the effectiveness of a European and trans-national approach for large and complex research projects in this area. Virtually all important industries and other development organisations in Europe working on PV cells and systems were involved in the programme. A large variety of components and systems making use of PV technology gave Europe an extensive competence for solar electricity. Many applications of PV were covered and prepared European industry for future markets.

The programme was successful. A large variety of high-performance PV technologies and systems became thus available in Europe ready to meet many applications. PV systems are viable in all European climates, they can be built environmentally and aesthetically attractive.

The Commission's services in Brussels when I was myself responsible of the Renewable Energies Division assumed general management responsibility throughout the programme.

The modules employed in the programme were representative of the state of the art at the time of construction in 1980. Modules were supplied by AEG and Siemens from Germany, France Photon and Photowatt in France, Pragma and Ansaldo in Italy, BP Solar in England and IDE/Fabricable in Belgium. The modules were qualified by the Commission's Joint Research Centre at Ispra in Italy.

Single and poly-crystalline silicon cells were employed. Module sizes ranged from 20 W to over 100 W and involved different encapsulation techniques. In general, European module development has benefitted considerably from this pilot programme.

The array: the internal array loss from mismatch of modules, cabling and diodes, a measure of quality of the array, could generally be kept as low as 2%.

The power conditioning: All plants used AC systems except 3, one of which was linked to a water electrolysis unit, another to a water disinfection unit, and still another to DC compression motors. Inverters were solid state with one exception (motor). Inverters were either designed for grid connection or stand-alone operation, some were suitable for both. Some systems used maximum power point trackers, others did not. The importance of managing power flow to ensure safe, reliable and efficient operation had been recognised and microprocessor control was widely used to this end. High efficiency of approximately 90% even at 10% load or 10% of maximum achievable insolation was achieved either in simple systems with one inverter or multiple switching units.

A full range of DC/AC inverters from 1 kVA to 75 kVA was developed by nine different European manufacturers; Jeumont Schneider, Aérospatiale and Leroy-Somer, France, ETCA, Belgium, Filectron and Italenergie in Italy, AEG and Siemens, in Germany and Holec in Holland.

Array support structures: Most development had to start from scratch. All structures were fixed, some allowing seasonal adjustment. Two arrays were roof-integrated, while the others were installed on flat roofs or on the ground. Design criteria depended on the wind load and hence from the site. Weight per square metre could be strongly reduced in comparison with typical structures employed hitherto. Many different designs and construction materials, metal or wood as it were, were used. Costs depended very much on the terrain but in several cases structure cost did not exceed 3 to 4% of the total plant cost, an important achievement.

Batteries: They were improved designs of lead-acid batteries provided by Varta in Germany and Oldham, UK.

Monitoring: Plants were generally unattended as they needed no maintenance. All plants employed the same standard monitoring system. Raw data were transmitted for exploitation to Ispra, acting as the Commission' monitoring centre for this programme.

For plant acceptance The Commission's Joint Research Centre in Ispra had developed a new device which measured the power of sub-arrays up to 10 kW. All plants were accepted this way.

Associated with the EU PV pilot plants of 1980 was an experimental setup of several small PV arrays at Adrano, near Catania in Sicily. As Adrano was the site of Europe's first solar CSP plant, a 1 MW solar tower plant, of which I had previously been myself the manager, this setup allowed the comparison between the values of CSP and solar PV.

The manager of the new PV project at Adrano on contract with the Commission was Dr Achille Taschini of ENEL, Milano.

The Adrano PV project had six plants installed for research, each with a capacity of 5 kW. Contractors with ENEL had been Ansaldo from Italy, Elf Aquitaine from Paris, MAN from Munich, Nukem from Germany, Solaris from Italy, Total from Paris. Hence, it was a demonstration of the new interest in PV of Europe's biggest energy operators.

(The initial plan was to have a project with four different types of flat-plate modules and two types of concentrator modules for comparison. The generated power was fed into the local grid of ENEL. 4 different types of PV solar cells were supposed to be used on the flat-plate arrays, Mono-and poly-crystalline silicon, and twice those of CdS/Cu_2S. The tracking arrays were to be equipped with mono-crystalline solar cells, one on a parabolic mirror concentrator with one-axis tracking and of a concentration rate of 25, the other a two-dimensional Fresnel array with two-axis tracking and a concentration rate of 40. The plants differed considerably from each other: the trapping surface needed for 1 kW varied from a minimum of 13 m^2/kW with silicon to 40 m^2/kW with CdS cell. So far the plan.)

Here is what was realised in terms of Adrano PV plant:

Four flat-plate silicon arrays of some 3 kW each were installed indeed.

- One from Ansaldo with poly-crystalline silicon. Tilt angle could be adjusted.
- One from Helios with mono-crystalline silicon. Periodic adjustment of the angle of inclination was possible.
- Two identical arrays on "Helioman", two-axis Sun tracking structures from MAN in Germany, with poly-crystalline solar cells from AEG. They had automatic tracking sensors.

For two-axis concentration, 46 spherical mirror concentrators were employed that had been designed by CISE in Italy. They had a concentration ratio of 600 suns. Each receiver has 30 GaAs cells with a heat exchanger to collect the thermal energy, too.

Eventually, a flat-plate array with a-Si cells was installed by Siemens. In the meantime, interest in CdS cells had already waned.

Generally, after the initial correction measures, all systems performed reasonably well, as expected.

At last are presented the residential PV systems realised as part of the EU Pilot Programme in the 1980s. They were among the first ones, today several millions of them are in operation in Europe and the world:

- **In Bramming, near Esbjerg,** Denmark, a PV powered house was built under contract with the EU Commission. The project had 5 kW of PV on a "standard Danish house". It was put into operation in August 1984. Prime contractor was Jutland Telephone. The house was built from wood. Great attention was put on building a compact but well-arranged house. It had to be a standard single-family house. The PV modules were architecturally integrated into the south slope. The system was grid connected but disposed of a battery, as this type of house was also designed for export, the stand-alone mode had to be demonstrated.

- **The "Herzog House in Munich".** 3 kW of PV was integrated in a modern building, grid connected, mostly built from glass. It was built in 1982 and put into operation a year later. It was sponsored by the Commission and some private firms. The architect was Th. Herzog. Fraunhofer ISE, Siemens, AEG, and others built the project. It had no storage batteries. Part of the PV panels were placed in the roof by replacing existing glass panels; it proved to be water tight. The building was optimised for passive solar heat gains. Modules were with poly-crystalline silicon cells and on double-glass, frameless. Modules were from AEG and from Siemens. **Operation has shown that PV would provide 90% of the demand of the house.**

- **PV House in Sulmona, Italy.** The project, also a design by Th. Herzog, was to show PV in a prefabricated house. It has been built but had no follow-up.

- **The Rappenecker Hof.** The 3.8 kW PV system is integrated in the building's roof. It is a farmhouse from the 17th century. Fraunhofer ISE claims that the system installed in 1987 was Europe's first inn supplied with PV. In later years ISE converted the project, once initiated by Jürgen Schmid from ISE, into a showcase for new energies by adding a small wind turbine and a fuel cell system.

- **Villa Guidini near Venice, Italy.** It is a 18th villa converted into a museum with a public garden. The 2.5 kW array from the company Helios was set up in the garden. Its angle of inclination could be adjusted. The system had a large battery provided by Fiamm of Montechio, Vicenza, and stayed grid connected.

- **Berlin PV Plant in a residential district.** A 10 kW system was installed in 1989 by BEWAG and the Institute of Prof. Hanitsch in Berlin. The PV arrays from AEG and Siemens were building integrated. It had a battery and was grid connected. It was used to provide supply to a heat pump, household appliances, and some lighting for the building, and serve as a charging station for batteries.

Chapter 9

The Later Years of the 20th Century: Big Oil, Japan, Africa, and the Developing Countries

9.1 The Oil Companies and PV

The San Diego Union-Tribune wrote in **2011**, "Big Oil bought and controlled the research and development of America's PV industry. Big Oil's main business is and has always been oil, gas, coal, and petrochemical profits first. As a result, the private sector stimulation was actually mergers and buyouts of smaller independent PV companies by big oil".

And in **1978**, the *Washington Post* had already remarked, "For years, consumer advocate **Ralph Nader** began pressing the government to underwrite the development of solar power as an alternative to the traditional forms of energy, dispensed by American industry. To his chagrin a group of Fortune 500 giants from Atlantic Richfield to Westinghouse has taken up the solar cause. The large corporations of which he is so wary have become the major beneficiaries of Nader's efforts. Already some of the small, innovative companies that Nader had hoped would develop are being bought up by the giants."

In 1973, the young PV industry for terrestrial applications did actually start in America and the oil industry became involved already right from the beginning. At that time, the

Solar Euphoria: The Rise of Photovoltaics to the Top
Wolfgang Palz
Copyright © 2023 Wolfgang Palz
ISBN 978-981-5129-00-7 (Hardcover), 978-1-003-43866-3 (eBook)
www.jennystanford.com

recognition of the need for alternative energies was widespread around the world and in America. With the first oil crisis in that year oil companies thought to look for alternatives to oil from the Middle East. It was an incentive for some to investigate PV's potential as a big energy provider in the future. The limitation of fossil resources was a real concern in society. And in 1973 came the concern of too much dependence on oil and gas from the Middle East. Fracking of oil in the United States and oil from Russia was not yet on the cards at that time. By now 50 years later, imports from Russia have much disappeared after the war in Ukraine and what remains is fracking, fracking in the United States and Canada on a big scale.

And there was the search for alternative energies that were sustainable. And PV is a climate respecting power per excellence. In this year 1973 we had organised the epochal UNESCO Congress "The Sun in the Service of Mankind". I remember how at a press conference in Paris in 1973 people asked already how much more time it would take for sea water to rise to dangerous levels.

And here is what happened since 1973. In April that year, **Elliott Berman** founded **Solar Power Corp**. with Exxon (it was closed in 1984). Later that year, Karl Böer formed **Solar Energy Systems** with Shell Oil to commercialise his CdS thin-film cells. In July 1973 was formed **Solarex**. For that one, it took until 1983 that it lost independence when Amoco, a subsidiary of American Oil Co. bought it.

In 1974, Mobil Oil with **Tyco** formed a joint PV company, to promote silicon cells from ribbons. It was sold to the German **ASE** in 1994.

In 1977, Atlantic Richfield bought Solar Technology International just established 2 years earlier that became **ARCO**. It was taken in 1995 by Shell after having transited through Siemens ownership.

In Europe BP took **Lucas** Energy systems in 1981. **BP Solar** is from 1983. BP merged with Amoco in 1998. In 1999 BP got Amoco-Solarex. BP left all PV activities from that time in 2011. In the Netherlands, Holecsol was taken by Shell in 1982. In 2007 Shell dumped many of its solar projects.

In 1980 in France **Photowatt** was created with Elf Aquitaine inheriting the silicon technology from France's first PV pioneer for terrestrial PV, the RTC in Caen, a subsidiary of Philips.

In 1976 was formed in El Paso **Photon Power** with support of the French oil company Total and some US companies. I was personally involved. The purpose of the company at that time was to develop the CdS solar cells to cover big areas of the deserts in the US south-west for power generation.

What happened in the late years of the last century in the PV industry in the United States when the PV business was still small, looks today a bit like games in a kindergarten. In Japan, Shell came in later in the PV industry in a marginal way. But PV leaders in those days did not come from the oil industry. A Japanese early leader was Sharp that offered PV for terrestrial application since 1963 already.

The Chinese leaders of the PV solar cell and module industry of today became big players on the global markets only when the new millennium had started. Their PV business in the hundreds of billions of dollars was getting comparable in size with the oil and gas business. Today PV is a giant on its own.

9.2 Global PV Leadership of Japan at the Turn of the Century

Japan lacks natural resources except the Sun and the renewables derived from it. Hence, the 1973 oil price shock triggered new interest in Japan's domestic energy, the Sun: A first "**Sunshine Project**" was decided.

But it was not the only energy. Nuclear energy started to benefit from the governments favours as well, with Fukushima as the result. But that is another story.

The Sunshine Project included electricity taxes, the creation of a special account to solar energy and the establishment of a new agency, **NEDO**. Sharp, Matsushita, Hitachi, Toshiba, NEC, and later Kyocera and Sanyo were programme participants.

Sharp was one of the world's first PV manufacturer, it had started in 1959 and offered commercial products since 1963. In 1976 the company introduced the pocket calculator. Sharp is still in the PV business until today.

In 1978, first PV systems in Japan were connected to the grid. It was a revolution giving a new PV market perspective to those who have already power at the expense to those who have not yet any.

One of the most important national incentive programmes was a government subsidy for residential PV systems which was in effect between 1994 and 2005. This policy was driven by the need to demonstrate the commercial success of the Sunshine Programmes as well as assist in the development of mass production capabilities of Japanese firms.

From 1990 to 1992 compact PV rooftop systems were developed by the industry. Sanyo started the PV programme on buildings in 1992. **It is important to note that new legislation was introduced to force the grids to buy back surplus electricity from the PV owners.**

Japan's industry became world leader for solar cell production in 1990; **still in 2005 globally half of all silicon modules came from Japan.**

In 2004, the year when its PV market was caught up by Germany's, Japan had 1.1 GW of PV installed. Between 1992 and the year 2000 Japan had 50,000 new PV rooftops. Most PV was installed on individual residences of environmentally conscious higher income families.

A few years later, things changed. Japan became from the world's number 1 module exporter to a PV module importer.

9.3   PV in Africa and the Developing Countries

Until 1995 most of the global PV business was off-grid, almost 100%. A famous first application when US companies became serious about PV in the early 1970s was PV for lighting on oilrigs in the Gulf of Mexico, then systems for the US Coast guard, communication devices. By the year 1995 still 20% of PV deployment in the world went to Africa. But global yearly deployment was still modest at 80 MW installed in the year. **As we saw here above, in 1992, PV on grid-connected applications had been started seriously. Today 99% of the Terawatt of PV installed worldwide is grid connected.**

In 2013, only 1.5% of global PV deployment went still to Africa, but it was mostly for the Republic of South Africa and the Arab countries in North Africa. **I called it in my publications the 1% scandal.** Since my friends and all the PV communities around the world had been looking around for early PV application since the 1970s, the obvious first choice was providing light and electricity to the billions of people who have none of those. **The UN Millennium Development Goals call for fighting the lack of electricity and light to 1600 million people.** Why should the peoples left behind in their development so far be encouraged to employ the same polluting and climate killing energies as we did, instead of going from the beginning to **solar energy,** and all the renewable energies. What were then the barriers to solar deployment in Black Africa?

First, the local utilities, with managers formed in the industrialised countries, funded by capitalists used to work with the stock market, wanted to do the same as their rich colleagues in the North: centralised power from big plants of hydro, oil. Fortunately, nuclear was technically speaking not possible, too big. Decentralised power: that used to be diesel sets, no matter that those are a lot more expensive than solar power, PV. And there are entrenched interest and business in the traditional energies.

I remember, in the mid-1970s when I worked in France, I came in contact with French people selling diesel sets in Africa. They got interested when I explained them PV. Another one sold educational TV sets powered by batteries for the kids in the village schools. He eagerly adopted PV, which helped him avoiding carrying all his heavy batteries to a central charging station.

Since the turn of the century, PV has become successful around the world, starting from Germany, as we'll see later. But there mainly the middle-class people finance the PV deployment; it is for their own interest. In Africa most people live from a few dollars a day. There is lack of money for private investment. The NGOs do not have enough power to help.

In those days, some 20 years ago, when Africa was beginning to be left behind in the solar boom, modern lighting with low consumption LED was not yet available and the indispensable batteries for stand-alone systems were lead-acid, and not lithium.

Examples of how electrification with PV has become recently successful are countries like India, where PV became booming just in a few years; why should Africa not take, in a few years from now, as well the route of PV, generalised power for everybody?

Chapter 10

PV Expert Meetings around the World, Political Monster Meetings in Support of the Renewables, New Energy Organisations

COP 21, or the UN Framework Convention on Climate Change (UN FCCC), held in Paris in 2015 was the most important reference yet for global reduction of GHG emissions and hence, limitation of Earth's temperature increase. It was highly political but did not deal much with solar PV. For PV there have been many others.

An **Agenda 21** was adopted in Rio in 1992 by 178 governments at a **World Conference on the Environment and Development.**

Followed in 2002, at Rio+10, the **Earth Summit in Johannesburg.** It was attended, with many, many others, by the German Chancellor. Our friend Hermann Scheer had convinced him to invite from there to another political conference in Bonn. And that one was specifically on the renewables:

It followed "**Bonn Renewables 2004**", with 3000 participants from 154 countries. Including myself this time. It called for a global transformation of our energy system and declared, "Renewable Energy will be a most important and widely available future energy system". And from there, my friends and I pushed things further:

Solar Euphoria: The Rise of Photovoltaics to the Top
Wolfgang Palz
Copyright © 2023 Wolfgang Palz
ISBN 978-981-5129-00-7 (Hardcover), 978-1-003-43866-3 (eBook)
www.jennystanford.com

Then followed the **International Renewable Energy Conference in Beijing** in 2005, at the Great Hall of the People. The same year the People's Congress of China adopted a law on the renewables. It adopted among other measures the FIT.

Followed in 2008 the "**WIREC**" **PV Congress** in Washington DC, with the participation of many US Officials. **President Bush gave a speech.** Eighty pledges were adopted.

The year 2008 was also the year of a global financial crisis that entailed a global crisis on the young PV industry—more on this later.

PV specialists had their own meeting in the United States since the 1960s already. The beginnings are told by Moe Forestieri in the book *Solar Power for the World*, which we published in 2014 (ISBN 978-981-4411-87-5). The first PVSC was held in April 1961 in a basement conference room in Washington DC. In 1962 followed the second in Washington DC. It was organised by the Interagency Advanced Power Group. In April 1963 "the 3rd Photovoltaic Specialist Conference" was held in the format used since then: it was held at Statler Hilton in Washington DC, sponsored by IEEE and NASA. A numbering system was actually only used after the 4th in 1964. Since 1967 IEEE is the only sponsor. The **50th Photovoltaic Specialist Conference PVSC-50** is being held in June 2023 in Puerto Rico.

Since 1973, the NASA Lewis Research Center organises the Space PV Research and Technology Conferences SPRAT. NASA Lewis at Cleveland Ohio on the Erie Sea was in charge of the PV development for applications in space. *I remember, when I first travelled to the United States in 1969 as the PV director at CNES in Paris and came to NASA Lewis, I was received like a prince: right from the director of the Center with photographer, etc.*

In Europe I organised myself the PV Conferences on behalf of the EU Commission since 1977. The first EU PVSEC took place in Luxemburg. For the second, we went to Berlin, at the "Oyster" building with a thousand participants. Prof. **Broser** from the Free University in Berlin, one of our hosts for the conference gave us the advice to start numbering our conference series, which we did. And like our friends from the IEEE in the United States, we changed the hosting city for the Conferences for each of them. After Berlin we went to Cannes in France, later came

Athens: a memorable conference opened at the amphitheatre just below the Akropolis with **Melina Mercouri** attending. The **40th EU PVSEC** is being held in 2023 in Lisbon by a commercial organiser.

Committee meeting of an IEEE PV conference in the US, the author being a member. The 50th conference is being held this year.

My friend, the late Prof. Hamakawa, a leading promoter of PV in Japan, started the first of a series of **Asian PV Conferences in Nagoya 1984.** Followed PVSEC 1986 in Beijing (that I also attended), later Tokyo, Sidney, etc. **PVSEC-34 is held in 2023 in Shenzhen, China.**

An offspring of the different PV Conferences are the World PV Conferences. The first took place 1994 in Hawaii after an agreement between the IEEE and the Asian Conference organisers that I had joined on behalf of the EU. The second was held in 1998 at the prestigious Hofburg in Vienna. It takes now regularly place every 4 years. The last one to date, the eighth was held in 2022 in Milan, Worth mentioning are also the global PV industrial exhibitions. They grew enormously with ten thousand of visitors as the PV business developed. **Intersolar** was first organised in 1991. It is regularly held in June each year in Munich.

The biggest PV event with 200,000 m^2 exhibition area is probably the **Snec** in Shanghai. It is also held regularly but had an interruption because of Covid. I had once the honour to share the opening ceremony, with trumpets and drums.

These industrial PV fairs used to have a conference associated with them.

Among the organisations nowadays important for observing and supporting the global PV markets, one can mention in particular the **International Energy Agency (IEA), International Renewable Energy Agency (IRENA) and REN21.**

The **IEA** was set up by Henry Kissinger from the United States right after the first oil price shock in 1974. Its offices are in Paris a step from the Eiffel Tower. It follows the development of all energies on the global markets, in particular that of oil. The IEA has international expert working groups about all energies, also on PV. In the last few years it took particular interest in PV.

IRENA in **Abu Dhabi**, exists officially since 2011 as an independent organisation, with Board, General Assembly, etc., organised like a UN agency. Its father was **Hermann Scheer.** The German government decided to organise it under the leadership of the late Hermann Scheer with the support of the TV journalist Franz Alt. Chancellor Merkel had given her ok. The founding conference took indeed place in Bonn at the old Parliament Building in January 2009; it was attended by over 60 state delegations and was chaired by Hermann Scheer. The conference to determine the IRENA headquarters and its first DG took place shortly after in June 2009 in Sharm el-Sheikh on the Sinai peninsula of Egypt. I attended it on behalf of the French delegation. Obviously, Hermann Scheer saw either the headquarters in Bonn or himself as DG. But under the pressure of the Arab delegation at the meeting, the German government delegation at Sharm el-Sheikh left Hermann in the rain. It was decided to have the headquarters in Abu Dhabi. A big disappointment for us. The first DG became a friend of mine, a lady from the French Government in Paris. But she was unhappy in the local environment and gave up shortly after. Right the first Assembly in 2011 was a monster meeting with a thousand participants, 50 ministers and 670 country delegates. A waste of money? Anyway, it cost a lot of it.

REN21 is also a totally German creation. It was set up in Paris by the Green J. Trittin when he was Environment Minister in Bonn in 2004. Why Paris? Why 100% funding from the German government? Trittin, from the Greens, did not particularly like Hermann Scheer from the SPD. So what Trittin wanted was an agency, but not one after the ideas of his colleague.

PART III

THE BIG BANG IN THE YEAR 2000

Chapter 11

New Political Regulations in Germany, New PV Industry in China

The big bang came on April 1 2000 when Germany adopted the EEG, the Renewable Energy Sources Act and the FIT. It was the merit of Hermann Scheer, MP. He reported on it in an article in my book *Solar Power for the World* (2014, p. 287). It was a long innovative construct that Hermann Scheer organised with the support of the Bundestag, the German Parliament, of which he was the leader for the renewable energies. And he was my best friend.

The year 1989 brought him the first political success. He authored a parliament resolution to prioritise the promotion of solar energy in research and development policy. The resolution passed unanimously by the Bundestag. One consequence was the 1000 Roof PV Programme by the federal government, a first systematic PV market launch programme.

The year 1990 saw his second success: in an effort mounted with colleagues, also some from the other major parties, the first feed-in tariff law for renewable energies was adopted by the Bundestag. It regulated guarantied grid access for electricity from renewable energies supplied by independent producers and guaranteed a feed-in price of between 75% and 90% of the underlying electricity price. The law was successful for the promotion of wind power, but for PV the law was not yet attractive enough. Up to the last minute the big utilities tried to

Solar Euphoria: The Rise of Photovoltaics to the Top
Wolfgang Palz
Copyright © 2023 Wolfgang Palz
ISBN 978-981-5129-00-7 (Hardcover), 978-1-003-43866-3 (eBook)
www.jennystanford.com

frustrate the legislation. It was later said that they knew how important this law was to be.

In 1993, Hermann Scheer presented a first **100,000 Roof PV Programme**. It seemed utopian even for the representatives of the still very small PV industry. The biggest German PV company at the time, Siemens Solar, declared the programme to be immoderate. But people started setting up solar clubs. Famous was the one **in Aachen** where **Wolf von Fabeck** called for a **"cost-covering price"**. The Aachen City Council took this up in 1994 and decided a local FIT of 1.80 DM/kWh, a lot of money at the time. In the course of the next 4 years, more than 40 towns and cities followed Aachen's example. These local initiatives made a crucial contribution to a public wave of support for PV. They created a base for a PV market. Without them it is unlikely that the small German PV market of the 1990s would have survived.

The year 1998 was a key year for the next steps. The outcome of the general election to the Bundestag brought a majority for the SPD, Hermann Scheer's party. He was re-elected. The SPD formed with the Greens a new government.

Just before, already in April 1998 Scheer succeeded in anchoring the 100,000 Roof Programme and the FIT in the SPD's election manifesto—while Greenpeace had asked for a 50,000 Roof Programme. It was then translated, after hard negotiations into the new government programme. Not losing time waiting that the federal budget become approved some months later, Scheer succeeded an immediate start by mobilising the new Finance Minister, Oskar Lafontaine, and the state-run KfW bank and an approval by the Bundestag.

So, the PV Roof Programme with the FIT was started on 1 January 1999. The programme was a zero-interest scheme with the difference between zero interest and a loan-market's interest rate being subsidised. To improve effectiveness of the scheme and PV uptake in the market one had to increase the Feed-in Tariff. To this end a new bill the **"EEG"**, a Renewable Energy Sources Act, was passed through the Bundestag. Before its final vote, a torpedo came from the EU Commission: Scheer called it a High-Noon Situation. Scheer won the final disputes and the EEG got its majority in the Parliament and became effective on 1 April 2000. The EEG referred to the "Aachen Model" mentioned before, paying a cost-covering price to the owner of the PV plant feeding electricity into the grid. The first EEG included a 45 cents/kWh price in addition to the benefit from the Roof Programme.

A week later, on April 7, the **EU Commission filed a lawsuit on the EEG** that was dismissed another 11 months later by the **European Court of Justice**. So, on March 13, 2001 the EEG for PV (and the other renewables) got its final GO AHEAD.

The Roof Programme together with the first EEG associated with it was expiring at the end of 2003. It was not renewed. A second EEG starting in **August 2004** remained as the only support mechanism. To compensate the lack from support of the Roof Programme, the tariff of the FIT was increased in the first EEG in 2004 from 45 cents/kWh to 57 cents/kWh. **All FIT payment is distributed evenly among the electricity consumers, typically a surcharge of 3.5 cent/kWh. It is not a tax. The FIT is not paid from the federal budget. The European court has decided that it is not a subsidy.**

The tariff paid to PV electricity producers in Germany is not linked to retail prices, it is fixed for 20 years. In the first EEG, rooftop PV capacity lower than 30 kW got the highest tariffs/kWh, larger PV systems got some 20% less. Tariffs were to be lowered each year. The digression rate was in the first years the same for each year. It was different for the different types of renewable electricity. For PV it was –9%, each year again. This reduction was justified by the learning curve by mass production.

The EEG confirmed the priority of PV's access to the grid.

The FIT was to be renewed every year; the first time in 2007.

Hermann Scheer our hero, father of the EEG with the FIT, passed away prematurely from a heart attack in 2010. He is unforgotten.

Germany started expanding its PV market since the EEG was finally in force with the "speed of light". In 2004, with the first GW of PV installed in the country, it passed Japan, the global PV leader by then thanks to its solar-roof programme from 1992 mentioned previously. Germany remained global PV leader in terms of yearly market installations until 2013. Thanks to further PV market acceleration from 2010 onward, Germany had installed **30 GW** of total capacity of PV in the country. The cost had been €200 billion.

In Germany PV reached grid parity by 2011/12.

Unlike "subsidy" for the installation of PV plants, paid per kW of power, the FIT comes per kWh of the electricity generated. It is more profitable in the sunny countries with more kWh produced per kW installed. In Europe, Spain and Italy are such countries. And these countries followed suit with their own FIT, among others.

In Spain the FIT was installed by a Royal Decree in March 2004. By 2006 my friends from the Research Centre **ITER on the island of Tenerife** organised the installation of one 13 MW PV plant in 2006 by private investment at which I participated (islands are PV-friendly markets in particular when they are not connected to a national grid, as is the case for the Canary islands). Until 2012 Spain had installed **5 GW**. By that time, the government cut all support. The 4 years until 2016 when PV installations resumed Spain's PV deployment came to a complete halt.

The 13 MW PV plant on the island of Tenerife, Spain. The author is one if its many owners.

In Italy, the FIT, called "Conto Energia" was installed in 2003. It led since 2007 to a high growth of the installation rate. When it was greatly reduced by 2013 the country had over **18 GW** of new PV installed—with a support of €6.7 billion in total.

By 2007 one counted 46 jurisdictions, many of them in the United States with a FIT policy for PV and all renewables in place. Its principle was adopted by over 65 nations. Many studies found that the FIT was the most effective for promotion of PV in particular in comparison with the Renewable Portfolio Standard (RPS) and the quota model.

Already between 1978 and 1988, global markets increased 35 times. In the first decade of this century the global PV market grew 32 times, on a much higher level as it were. **But at that time PV was expensive. It was the most expensive on the global electricity markets in purely economic terms. Not like today where it is the cheapest. The European countries Germany, Italy, Spain and others and their grid clients paid for the surcharge to trigger the start of mass production. It was not paid from state budgets.**

For the PV industry these FITs and the new favourable and profitable markets came as a surprise. Industrial production capacities of cells and modules had to be built up, mostly from scratch. First thing to do was automation of production chains. Germany with its strong mechanical industry came to the forefront. Heavy investments had to be mobilised. In Germany, for instance, the rumour went that everybody who could spell "photovoltaics" opened a company. By 2012, thousands of new PV companies had established themselves, manufacturers and installers. 110,000 jobs had been created in Germany. Billions of euros were invested in new firms, many of them stock market darlings. Just in 2010, €20 billion of investment in PV was accounted in Germany.

In a liberalised market, PV importers got their chance. From Japan came the PV leader Sharp and others. **But the PV wake-up call was, in particular, received in China. Immediately since the year 2000, new PV companies were formed in China. Their products were competitive in quality and price. But China did not produce for the PV home market, as by that time it**

did not have any. Most of Chinese PV went to Germany and the European markets.

In return, Germany could sell the PV production machinery to China.

In 2009, Global PV solar cell production reached 10,660 MW, exceeding **10 GW for the first time**. Almost half of it was manufactured in China and Taiwan. In that year hundreds of PV companies existed in China.

The lion's share of 8265 MW was with crystalline silicon cells. CdTe cells exceeded 1 GW for the first time. The 10 largest solar cell producers in 2009 were First Solar, then Suntech in China, Sharp in Japan, Q-cells (at that time in Germany), Yingly, JA Solar, Kyocera, Trina Solar, all Chinese, SunPower, from the United States, and Gintech. More than half of production went to Germany, followed by Italy, Japan and others.

Global turnover in 2009 was, like in the previous year $30 billion. For the first time production costs of less than $1/watt were achieved for CdTe modules, and only slightly more for silicon cells. Strong competition in the global markets produced price decreases of 50% in one year.

For complete PV systems prices came down to $3/watt in 2009.

Chapter 12

2012, PV's Year of Industrial and Social Disaster, Germany Giving Up Its Leadership

Remember **2008, the Lehman Brothers.** The US bank's failure led the world to a financial crisis. Frau Merkel's Germany went into recession. The credit crunch in 2008 resulted in a large number of PV projects being cancelled or delayed. Of the 7.5 GW of global PV production in the year 2 GW remained unsold. Solar stocks were down 76%, twice the decline of the general market. Module prices have been cut in half. The world's solar industry was in full swing, but with Lehman it crashed down to Earth.

Thanks to the tremendous growth of the PV markets in the preceding years the silicon market for PV had already become bigger than the market of silicon for computer chips. Thin-film PV cells and modules got great attention, more than 140 institutions and companies were active in the new industry. Its market share grew much slower than expected previously. A bit like perovskite solar cells by now.

The PV business has changed in 2009 to a buyer's market. Since January that year, prices of modules have fallen by 30% to 40% as the result of overcapacity and unexpectedly sharp fall in demand. Chinese modules, in particular, became cheaper. End users welcomed a new situation.

Solar Euphoria: The Rise of Photovoltaics to the Top
Wolfgang Palz
Copyright © 2023 Wolfgang Palz
ISBN 978-981-5129-00-7 (Hardcover), 978-1-003-43866-3 (eBook)
www.jennystanford.com

The financial crisis did not last for long. **In 2010** Germany's economy rebounded strongly. The PV module producers must have taken then the wrong conclusions, expecting a recovery of the PV market as well. And it did: **the German PV market reached, for the first time, the 7 GW level of PV installations in one year.**

But this in turn made the paymaster for the FIT unhappy, the German electricity end users that pay the EEG apportionment. From €3.5 billion in 2004 it went up to €12.7 billion in 2010, also a new record; 3.5 cents/kWh for each power consumer. The government reacted by **reducing the FIT by 25%** in the single year 2010. It was against the –9% that had to be expected.

The PV module producers had by then their production lines running. They reacted by bringing again 7 GW to the German market—no matter the torpedoes from the government. **In 2011, Germany installed again 7 GW of PV.** It was certainly without a profit. The EEG apportionment went up to €13.5 billion. The government reacted by reducing the FIT by another 18%, twice the expected reduction of –9%.

Whatever the FIT, PV production lines were running and had to get their cells and modules to the market. **In 2012, again 7 GW of PV was installed in Germany: this time with a hefty loss to the module manufacturers. From deep red they had to give up their business. It followed the doom of the global PV industry.** A lot of money was lost at that time by the investors in PV.

Germany's own new PV industry lost half of the 110,000 jobs that had been created in the boom. Later in 2015 only 31,000 jobs had been left.

Almost all of German cell and module manufacturers went bankrupt in this disaster:

- Q-Cells, the world's biggest module producer came under Korean umbrella
- Conergy
- Solon
- Centrotherm
- Sunways

- Schüko
- Scheuten that had taken over Shell-Gelsenkirchen
- Würth, a specialist for CIS, a technology it had adopted from ZSW in Stuttgart
- First Solar, the CdTe specialist quit Germany at the time

Bosch had bought half of Ersol in 2008 and Aleo in 2009, gave PV up in 2012 and had to write off a loss of €1 billion.

Siemens Solar gave up after a total loss of €0.8 billion.

Schott Solar gave finally up in 2013. Its origins go back to 1958 to the famous AEG-Telefunken, later Deutsche Aerospace (DASA). In 1979 it had been created RWE-Nukem, a thin-film module producer. In the same year, MBB, the aerospace company in Munich, and Total Energy set up Phototronics, specialising in amorphous silicon modules. In 1994, all the three, DASA, Phototronics, and Nukem became a joint company, the new ASE under the roof of RWE, Germany's nuclear champion. In 2002 had entered Schott the scene, the German glass manufacturer. 3 years later, Schott, RWE Solar had become a wholly owned subsidiary of Schott. In 2012, the firm had given up its silicon activities, a year later those in the thin films.

Solar World went bankrupt several times. Created from scratch in 1988 in Bonn, it had its first crunch in 2013, only saved by Qatari investors. It had lost 99% of its stock market value. It went down the drain again in 2017 after having made life difficult for Chinese PV importers through political initiatives for duties in Europe and the United States. In March 2018, the dragon of bankruptcy set down a third time on Solar World.

In the United States, the company Evergreen, a specialist of silicon ribbon technology went down the drain. So did ECD, the promoter of a-Si on flexible substrates. The specialist of organic modules, Konarka took the same route out. A scandal was the insolvency of Solyndra as it had benefitted from support by the administration shortly before. Later in 2016, Sun Edison also perished. As mentioned previously, BP gave up its PV interests in 2011.

In Japan, the traditional leaders of the global solar industry, companies like Sharp, Kyocera, Showa Shell, Panasonic, and Solar Frontier felt also this tsunami from the European PV earthquake, but were never affected in their existence. But Japan that was a PV exporting nation until 2012 became a net importer of modules. In 2018, Kyocera's sales went down by 40% year on year and it made a loss of $616 million. Japan had adopted a national FIT in 2012 and stopped the industrial decline in the country. The market exploded and came slowly down in the following years.

In China, Suntech Power perished, too.

The situation became really absurd. Germany, Italy, Spain and all the other FIT pioneers have triggered the global PV market development—at the expense of the electricity consumers in these countries. They paid the bill to make PV cheap. But after all these industrial problems where again a big bill of billions of euros had disappeared, Europeans sidestepped from PV investments. **Germany installed less than 2 GW in 2016, Europe hardly 7 GW and China installed 24 GW in 6 months.** The first year that China became world leader for total PV capacity installed before Germany. China's yearly PV installations had passed Germany's already in 2013. China has adopted a national FIT in 2010. Since 2013 its domestic market started exploding. But it took until 2017 that China became world leader of total PV capacity installed.

The United Kingdom is a special case of European PV. It had adopted the FIT in 2008, becoming effective 2 years later. And the government did the right thing. It reduced the FIT down to a reasonable level from the beginning and had not to make it prohibitively low when the market took off. At some stage, the UK got the largest PV market in Europe despite the countries not so sunny climate. In 2019 it had a respectable 13 GW of PV installed, as much as France.

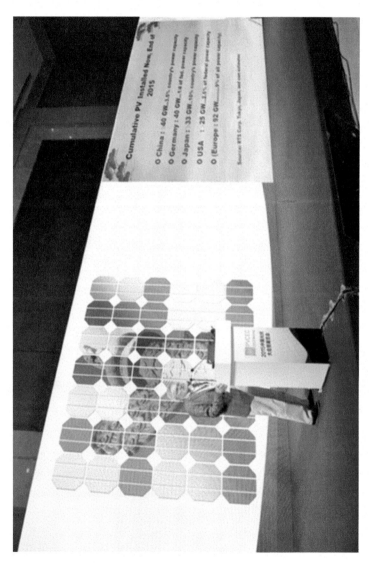

The author's speech in Beijing 2015 when China had just reached Germany's market in size. Since then, China's PV market has exploded to over 400 GW today, leading the world.

Chapter 13

PV Markets up to Now: A Summary

As reported previously, the world followed an explosive growth of PV, and the whole world took part in the party, for a clean and climate-friendly energy for everybody at low cost:

- 1 GW of PV achieved in the year 2000, covering an area of 5 km^2
- 10 GW achieved in 2009
- 100 GW achieved in 2012
- 500 GW achieved in 2018
- 1000 GW achieved in 2022
- 1500 GW achieved in 2024

The world's PV game leaders were Japan until 2004, then Germany until 2016, and since 2017, with flying colours, China.

In 2018, 100 GW was newly installed, almost half of it in China. By 2022, it increased to over 210 GW of new PV. China had end 2022 in total 392.61 GW installed, the double of 2018 (figures from the Chinese NEA).

The total 500 GW in operation globally in 2018 generated 600 TWh of electricity, 1.8% of consumption. Countries with the highest share of national generation in 2018 were Germany, Italy, and Japan.

Solar Euphoria: The Rise of Photovoltaics to the Top
Wolfgang Palz
Copyright © 2023 Wolfgang Palz
ISBN 978-981-5129-00-7 (Hardcover), 978-1-003-43866-3 (eBook)
www.jennystanford.com

More details about the PV markets later until 2023 have already been given in Section 2.2.

Following Bloomberg, BNEF, $130 billion was spent on PV in 2018. It was less than the previous years as the cost of systems had declined from year to year. Only $32 billion of the $130 billion went on the global module market, a large majority was spent on the Balance Of System BOS and other soft costs.

In 2022 half of all the world's power investments went on PV, half for utility-scale PV, the other half on distributed PV.

PPVX, the stock market index for PV, stood in early 2023 at some 3500 points from 900 in 2018. It increased strongly in 2020 and stood flat since then. The total market value stood at €135 billion in 2023.

PV has for several years now been the cheapest of all electricity sources, from fossil, nuclear, and to a short margin the electricity from wind power. But there will be no 100% of PV in the future. We discussed previously how the intermittency problem can be overcome—also by hydrogen and fuels derived from it.

Chapter 14

Green Hydrogen, the Follow-Up of Solar, Wind, and Hydro Power

Since the year 2000, solar energy and the other renewables have enjoyed a tremendous development, leading to a much more sustainable electricity supply. But next to global demand for electricity, the most noble among all forms of energy, there is the consumption of fuels that are not yet renewable, except for a small contribution from biomass (in France, a European frontrunner on biogas the target is to increase its share from 1% today to 10% of supply in 2028). **Today, 80% of global energy demand is still met from fossil fuels, oil, natural gas, and the different forms of coal.**

To replace those, there is "**green hydrogen**" on offer for a carbon-free economy. We are going to need tremendous amounts of it, including its derivates methanol, methane, ammonia, and others. The universe has developed from hydrogen and it is still full of it. Most of it on Earth is bound in water. Solar hydrogen, or green hydrogen when developed with other types of renewable electricity, hydro, wind power, is produced back from the water by electrolysis. Some politicians are backing as well electrolysis by nuclear power to produce the "**blue hydrogen**". Cost and efficiency of electrolysis are still poor today. Up to 80% of the energy are lost in the transformation process of the green electricity. Currently 1 kg of "grey hydrogen" costs €1.5. "Green hydrogen" comes €5.5.

Solar Euphoria: The Rise of Photovoltaics to the Top
Wolfgang Palz
Copyright © 2023 Wolfgang Palz
ISBN 978-981-5129-00-7 (Hardcover), 978-1-003-43866-3 (eBook)
www.jennystanford.com

There is the "**grey hydrogen**" generated from methane. The process is associated with tonnes of CO_2 for a single tonne of hydrogen produced. This traditional hydrogen, the grey hydrogen, is used as raw material for fertilisers, cement, etc., but not for energy purposes. **Yearly production is 70 million tonnes**, a lot. It will take hundreds of gigawatts of PV and renewable power to generate green hydrogen at this order of magnitude in the future. Linde at Leuna in Germany is a major producer of the grey hydrogen today.

A pioneer for green hydrogen was my colleague from the EU Commission, the late **Joachim Gretz**. He organised in 1990 the **100 MW Euro-Quebec Hydro-Hydrogen Project.** As the Province of Quebec in Canada owns large hydro resources and hydro-electricity, a renewable resource, the idea was developed to generate the hydrogen via electrolysis on site and ship it to Germany. Gretz, who was originally a citizen of Hamburg, had arranged with the local mayor to reserve a site at the harbour there for downloading the hydrogen from Quebec. For an innovative project of this size, the hurdles were too high, however. The project did not get from the ground and its supporter J. Gretz passed away in 2013. The world was happy with its fossil fuels.

A political change came with the UN Paris Agreement on Climate Change signed in 2016.

A factual change came with the new maturity of solar and wind power. Since 2012 and progressively later, they became the cheapest on the market: a new availability to go from there to water electrolysis for **"cheap hydrogen".**

But **solar hydrogen** is a change of paradigm. A whole infrastructure has to be revolutionised. After all accounts green hydrogen was three times more expensive than classical fuels. Costs were for the time being prohibitive. One had to work down a whole value chain to make green hydrogen cost competitive; similar to PV when the world started its development, decades ago. What was achieved for PV must become possible for solar hydrogen in the future: a commercially viable product.

A great hype for green hydrogen came with the **Paris Treaty.** Numerous initiatives and projects came about across the whole world:

- A Hydrogen Council was created in 2017.

would be generated with wind power, not hydro. And even better: an agreement was reached to install by 2025 at a former coal power plant in Hamburg, a 100 MW electrolyser with local wind power. Partners are Shell, Vattenfall, and Mitsubishi Heavy Industries.

Airbus is also supported from the German side to investigate future hydrogen airplanes.

Interesting to note is an initiative in Berlin to develop hydrogen for residences. The firm **Home Power Solutions (HPS)** proposes in a demonstration building there the combination of a PV plant on the roof, a small battery, hydrogen production in an electrolyser and its storage in high-pressure bottles, the whole in a small cabinet. The company has thousands of requests.

Europe is not alone with its interest in green hydrogen.

In the United States, the federal administration's DOE has a Hydrogen and Fuel Cell Technology Office. Among the many projects in the country, we may just mention a project in South Texas, one of the biggest. It is set up by **Green Hydrogen International (GHI)** aiming at the production of 2.5 million tonnes of hydrogen from 60 GW of local PV and wind power, with the use of a local salt cavern for storage. The possible employment of the hydrogen is its use as a rocket fuel.

American Airlines announced strategic investment in **Universal Hydrogen Co.** a company active in green hydrogen distribution and logistics networks for aviation.

China has in Inner Mongolia a little project to produce some 70,000 tonnes of hydrogen with 1.9 GW of PV and 400 MW of wind power.

In Kasachstan is planned by the firm **Svevind** a 45 GW of PV and wind power plant to produce 3 million tonnes of hydrogen per year. It may be used inter alia for the production of ammonia, steel or aluminium.

Another 50 GW PV and wind power plant for hydrogen generation was announced in 2021 by a consortium in **Western Australia; the Asian Green Energy Hub (AREH)** would produce 3.5 million tonnes of hydrogen per year or 20 million tonnes of ammonia. It would cost $52 billion, but its start is planned for 2028 only.

And at last comes the cavalry: British BP and French Total made big announcement in late 2022.

BP is to take a stake in the AREH for hydrogen in Australia as mentioned above.

Total Energies goes into an Indian venture with Adani aiming to produce green hydrogen with solar and wind power, at the beginning 1 million tonnes per year...

The End

semiconductor-grade 93
ultra-pure 97
silicon cells 6, 9–10, 94, 102, 108, 130, 148
crystalline 116, 148
poly-crystalline 124, 127
silicon crystals, ultra-pure 94
silicon market 149
silicon nitrite 8
silicon oxide 6
silicon solar cell 6, 94, 101
basic features of 7, 9
silicon wafers 7
solar cell producers 148
solar cells 7, 63, 93–95, 97–98, 102, 116
first industrial 97
first selenium 93
first silicon 94, 96
mono-crystalline 94, 126
perovskite 149
poly-crystalline 126
selenium 93
thin-film 4, 95, 108, 115
solar CSP 22
first 126
solar energy 20, 53, 85, 103, 108, 110–111, 114–116, 130–131, 133, 157
large-scale utilisation of 101–102
promotion of 108, 143
solar farms 32, 53
solar heating 108, 110
solar irradiance 83
solar irradiation 93
direct 4
strong 4
solar lanterns 63
solar parks 64
solar power 16, 23, 34, 51, 101, 103–104, 114, 129, 133, 136, 143

solar radiation 16, 85–86, 91
Solar Radiation Atlas 115
solar tariffs 28, 63
solar water pumps 63
solid silicon blocks 7
South Africa 22, 70, 133
Southeast Asia 65
Spain 4, 11, 16, 21, 25, 36–37, 39–40, 49–51, 70, 73, 146–147, 152, 160
steel, green 54, 160
storage batteries 127
Switzerland 7, 36, 55–56, 98
electricity prices 56

tax credits 29–30
technology
half-cell 10–11
silicon ribbon 151
Thailand 12, 32, 67
thermal solar collectors 108
thermal solar energy 110
thin-film cells 94, 116
thin-film module producers 32, 151
thin-film PV cells 149
Turkey 21, 26, 36, 39, 54–55

UNESCO 107, 109
United Kingdom 16, 21, 36–37, 39, 52–53, 110, 123, 125, 152
United States 4–7, 9, 12, 23, 26–34, 36, 38, 61, 75, 82–85, 93–95, 97, 99, 101–103, 105–106, 108–109, 111, 117–118, 130–131, 136, 138, 147–148, 151, 161

California 10, 20, 25, 27, 31, 33, 35, 73, 98, 104–106, 113
electricity generation 30
grids 26
market 29
NASA 98, 101, 103, 119, 136
Solar Energy Manufacturing for America (SEMA) 30
solar industry 30
Texas 28, 31, 33, 35, 105, 121
US Energy Information Administration (EIA) 26, 28, 31, 35
US National Oceanic and Atmospheric Administration (NOAA) 73
US National Renewable Energy Laboratory (NREL) 12–13, 34, 84, 94, 102
utility-scale PV 24, 28, 156

Vietnam 12, 21, 32, 65–66

wind
  offshore 24, 54
  onshore 24
wind power 4, 23–24, 26, 31, 34–35, 37, 39–40, 44, 57–59, 61–62, 64–66, 68–69, 81–85, 87, 110, 114, 143, 156–159, 161–162
World Health Organization (WHO) 50, 74
World Meteorological Organization (WMO) 74, 76
World PV Conferences 138

zero-emission urban development 116